# Tinbergen's Legacy

Nobel laureate Niko Tinbergen laid the foundations for the scientific study of animal behavior with his work on causation, development, function and evolution. In this book, an international cast of leading animal biologists reflect on the enduring significance of Tinbergen's groundbreaking proposals for modern behavioral biology. It includes a reprint of Tinbergen's original article on the famous 'four whys' and a contemporary introduction, after which each of the four questions are discussed in the light of contemporary evidence. There is also a discussion of the wider significance of recent trends in evolutionary psychology and neuroecology to integrate the 'four whys'. With a foreword by one of Tinbergen's most prominent pupils, Aubrey Manning, this wide-ranging book demonstrates that Tinbergen's views on animal behavior are crucial for modern behavioral biology. It will appeal to graduate students and researchers in animal behavior, behavioral ecology and evolutionary biology.

JOHAN J. BOLHUIS is professor of Behavioral Biology at the University of Utrecht, The Netherlands. His main research interests are in the behavioral, cognitive and neural mechanisms of learning, memory and development, in particular song learning in birds. He has been an editor of the leading journal *Animal Behaviour* and President of the Royal Dutch Zoological Society.

SIMON VERHULST is assistant professor of Behavioral Biology at the University of Groningen, The Netherlands. His research interests have included ecological immunology, avian life history evolution and avian energetics, all with the aim to study the interplay between function and mechanism of behavior. In 2004 he was elected as member of 'The Young Academy', the junior section of the Dutch Royal Society. He has been an Associate Editor for the *Journal of Animal Ecology* since 2007.

# Tinbergen's Legacy

## Function and Mechanism in Behavioral Biology

Edited by

JOHAN J. BOLHUIS
*Utrecht University,*
*the Netherlands*

SIMON VERHULST
*University of Groningen,*
*the Netherlands*

CAMBRIDGE UNIVERSITY PRESS
Cambridge, New York, Melbourne, Madrid, Cape Town, Singapore,
São Paulo, Delhi

Cambridge University Press
The Edinburgh Building, Cambridge CB2 8RU, UK

Published in the United States of America by
Cambridge University Press, New York

www.cambridge.org
Information on this title: www.cambridge.org/9780521874786

First published 2009

Printed in the United Kingdom at the University Press, Cambridge

This book is based on and expanded from a journal issue publication,
originally published by Koninklijke Brill NV, Leiden, the Netherlands
(http://www.brill.nl). This issue is entitled: *Animal Biology* (ISSN 1570-7555)
volume 55, no. 4 (2005) "Evolution, Function, Development and Causation:
Tinbergen's Four Questions and Contemporary Behavioural Biology", and
was guest-edited by Johan J. Bolhuis and Simon Verhulst. Brill owns the
copyright in all parts of the above special issue, with the exception of
Niko Tinbergen's paper included therein. Brill has provided Cambridge
University Press with a non-exclusive licence to reproduce the content
of this issue of *Animal Biology*. Where applicable in this book, proper
reference to the original articles in *Animal Biology* has been made.

*A catalog record for this publication is available from the British Library*

*Library of Congress Cataloging in Publication data*
Tinbergen's legacy : function and mechanism in behavioral biology / edited
by Johan J. Bolhuis, Simon Verhulst.
p. ; cm.
Includes bibliographical references and index.
ISBN 978-0-521-87478-6
1. Animal behavior – Research. 2. Tinbergen, Niko, 1907–1988.
I. Bolhuis, Johan J. II. Verhulst, Simon.
[DNLM: 1. Tinbergen, Niko, 1907–1988. 2. Ethology – history. 3. Behavior,
Animal. 4. Psychology, Comparative – history. QL 750.5 T587 2008]
QL751.T552 2008
591.5–dc22
2008021974

ISBN 978-0-521-87478-6 hardback
ISBN 978-0-521-69755-2 paperback

# Contents

*Contributors*                                          *page* vii
*Foreword by Aubrey Manning*                                    ix
*Preface*                                                      xxi

1.  On aims and methods of ethology
    N. TINBERGEN                                                1

2.  Tinbergen's four questions and contemporary
    behavioral biology
    JERRY A. HOGAN AND JOHAN J. BOLHUIS                        25

3.  Causation: the study of behavioral mechanisms
    JERRY A. HOGAN                                            35

4.  Tinbergen's fourth question, ontogeny: sexual and
    individual differentiation
    DAVID CREWS AND TON GROOTHUIS                             54

5.  The development of behavior: trends since Tinbergen
    (1963)
    JERRY A. HOGAN AND JOHAN J. BOLHUIS                       82

6.  The study of function in behavioral ecology
    INNES CUTHILL                                            107

7.  The evolution of behavior, and integrating it towards
    a complete and correct understanding of behavioral
    biology
    MICHAEL J. RYAN                                          127

8.  Do ideas about function help in the study of causation?
    DAVID F. SHERRY                                          147

9.   Function and mechanism in neuroecology: looking
     for clues
     JOHAN J. BOLHUIS                                    163


     *References*                                        197
     *Index*                                             228

# Contributors

**Johan J. Bolhuis**
Behavioural Biology, Utrecht University, Padualaan 8, 3584 CH Utrecht, the Netherlands.

**David Crews**
Section of Integrative Biology, University of Texas at Austin, Austin, TX 78712, USA.

**Innes C. Cuthill**
Centre for Behavioural Biology, School of Biological Sciences, University of Bristol, Woodland Road, Bristol BS8 1UG, UK.

**Ton G. G. Groothuis**
Behavioural Biology, University of Groningen, PO Box 14, 9750 AA, Haren, the Netherlands.

**Jerry A. Hogan**
Department of Psychology, University of Toronto, Toronto, Canada M5S 3G3.

**Aubrey Manning**
Institute of Evolutionary Biology, School of Biological Sciences, University of Edinburgh, Ashworth Laboratories, West Mains Road, Edinburgh, EH9 3JT, Scotland, UK.

**Michael J. Ryan**
Section of Integrative Biology C0930, University of Texas, Austin TX 78712, USA.

**David F. Sherry**
Department of Psychology & Program in Neuroscience,
University of Western Ontario, London, Ontario, Canada N6A 5C2.

**Simon Verhulst**
Behavioural Biology, University of Groningen, PO Box 14, 9750
AA, Haren, the Netherlands.

## Foreword: Four decades on from the "four questions"

As a former student of Niko Tinbergen, it is a pleasure to introduce this collection of papers written to commemorate the 40th anniversary of his classic paper on the "four questions" (Tinbergen, 1963, this volume). First a little history, because we need to remember the context in which Tinbergen came to write this memorable paper. It was designated as a salute to Konrad Lorenz, old friend and colleague, for his 60th Birthday, but its timing and content signified much more than this. During the late 1950s and early 1960s ethology was evolving rapidly and going through some turmoil. I suppose the original framework of "classical ethology" as it appeared to most of us at the time was encapsulated in Lorenz's 1950 Society for Experimental Biology Symposium paper and Niko's *The Study of Instinct* (1951). The number of people calling themselves ethologists was increasing rapidly especially in Germany, the Netherlands and Britain and many of them had close connections with Lorenz's and Tinbergen's groups. We were aware that this "new" approach to the study of animal behavior had plenty of antecedents both in Europe and in the USA. Tinbergen himself lists Whitman, Heinroth, and Verwey; nevertheless, I don't think we can be blamed for regarding it as new. After the Second World War, animal behavior was emerging from decades when American experimental psychology held sway – everyone should dip into Munn's (1950) heroic text to get some flavor of that influence – and Lorenz and Tinbergen seemed to signal a new era. They certainly came as a breath of fresh air to all zoologists and particularly to field workers. I joined Tinbergen's group at Oxford just 2 years after he arrived in Britain and so was privileged to be part of this wave of new work. We were definitely proud of our position; we called him the "Maestro" (in this we followed Medawar as Desmond Morris (1979) has pointed out) and we called ourselves the "Hard Core!"

International ethology conferences were held every couple of years and, being quite small, they certainly had something of a family feeling. We seemed to manage quite well within a fairly proscribed theoretical framework with species-specific behavior very much at its heart. Each species exhibited a repertoire of behavior patterns which were "innate," i.e., whose development was largely under genetic control, and whose performance was under the control of particular motivational states and sets of external stimuli. These latter were matched by "innate releasing mechanisms" in animals, which responded preferentially only to certain aspects (sign stimuli) in the external world – often sign stimuli were specially evolved structures or displays from conspecifics ("releasers"). Both Lorenz and Tinbergen had provided models for the organization of instinctive behavior. Lorenz's famous "psychohydraulic" model was much discussed – it modeled field observations during the breeding season rather well. Tinbergen developed a hierarchical model which, although it used terms like "impulses" with a more physiological sound than Lorenz's reservoirs and spring loaded valves, was really just a set of "black boxes" endowed with certain properties and connections.

The publication of Lehrman's (1953) critique of Lorenz's behavior theory produced major ripples in this rather small pond. To change the metaphor, it began to open up a rift between two major groupings of ethologists. The German group – *sensu lato* – reacted most strongly, regarding Lehrman's criticisms as an almost total rejection of the reality of innate behavior and exhibiting an obsession with learning processes. There can be no doubt that Lehrman did go rather over the top in proposing that some behavior which appeared fully formed, as it were, at the first performance, could be the result of hitherto unconsidered earlier experience. For example, Lorenz was particularly infuriated by Lehrman's citing work suggesting that the rhythmic movements of the head of chicken embryos in the egg induced by the beating of its heart were the origin of the pecking movements that young chicks exhibited upon hatching. Lorenz, reasonably enough, pointed out that, whilst all embryo birds were subjected to the same passive head movements in the egg, most did not exhibit pecking movements upon hatching, but gaped upwards in order to solicit feeding from their parents! By contrast the "English-speaking ethologists" (this was Lorenz's term and, I think, must include the Dutch!) although not failing to make some strong challenges to Lehrman, were more positive. They latched on to his key message; that a catch-all phrase like "innate" was in danger of making us ignore, or at least de-emphasize, the way behavior *develops*.

This will seem a modest conclusion and it is hard to believe it became such a contentious issue. Lorenz always asserted that his group actively studied development, e.g., the work on sexual imprinting in ducks, but in his writings and at meetings he always seemed to emphasize the contrast between innate and acquired components of behavior. Further, "acquired" seemed to equal "learnt." Great emphasis was put on so-called "Kaspar Hauser" experiments in which animals were reared in isolation of various degrees and later observed in their normal environment. Very often they performed remarkably well and so one could deduce that conventional learning and other types of experience were *not* required for normal development of this particular behavior – it is "innate." All too often this label was as far as it went. In fact, of course, such a result can best serve as a starting point for a study of development, i.e., to discover what type of experience *is* required and by what developmental processes does it make its mark?

Looking back on it, I find Lorenz's attitude unfathomable. The importance of his contributions to the new ethology were never in doubt. Perhaps he was just not interested in development as we, the English-speakers saw it. He certainly continued to believe that innate and acquired behavior came in distinct "packages" and it was almost as if he felt we devalued the beauty and power of innate behavior by pursuing its origins in the individual.

This was the background into which Tinbergen launched his four questions paper. He put it into the *Zeitschrift* because this was the main journal of the German-speaking ethologists at that time. It is now possible to understand his feelings in some detail. We have Kruuk's (2003) perceptive and certainly not uncritical biography and also Burkhardt's (2005) admirable history of ethology. Burkhardt has analyzed the correspondence between Tinbergen and Lorenz around this time which reveals that Tinbergen was seriously concerned by the rift and misunderstandings which were becoming so prominent. He was also well aware that Lorenz would not like some of the points he felt must be made. He says: "I have not hesitated to give personal views even at the risk of being considered rash or provocative." It was the attitude concerning development which was perhaps the most sensitive, but there were other issues from what we might call classical ethology which Tinbergen felt needed attention. I believe this remains the case today.

There has been quite a lot of navel-gazing by ethologists worrying about whether ethology has been overtaken by the emergence of other approaches or even perhaps, is now extinct! (See several essays in Bateson and Klopfer, 1989, for example.) Wilson's (1975) famous diagram in his *Sociobiology* text comes to mind where the giant amoebas of neurobiology and sociobiology are engulfing the last vestiges of ethology! I believe all this is a complete misinterpretation of what has happened. The great bulk of the classical ethological model outlined above is little considered now, similarly neither are Hull's behavior model or Skinner's. Ethology's enormous contribution was to reawaken the serious study of any animal's behavior, taking into account the selection pressures imposed by the environment in which it has evolved. In this sense it continues to dominate animal behavior studies; we are all ethologists now.

I have yet to meet anyone who does not accept that Tinbergen's four questions for the study of behavior – function, evolution, causation, and development – are the right ones. Of course, he discusses them with a very particular agenda in mind, which was to ensure progress on all four fronts with each paying close attention to the others. Further, he saw each question as applying to all levels of analysis, from physiology to population ecology. It is clear that, taking all this on board without any theoretical difficulty, progress in practice has been very uneven and a number of issues which Tinbergen discussed have effectively been left on the sidelines. Marian Dawkins (1989) referred to an ethological beast with four legs, one of whose legs is considerably longer than the others; this is "function."

Much of this is inevitable and simply relates to accessibility and the way science progresses. I have often reflected on Medawar's wise words in his *The Art of the Soluble* (1967).

> "Scientists study the most important problems they think they can solve. It is, after all, their professional business to solve problems, not merely to grapple with them. The spectacle of a scientist locked in combat with the forces of ignorance is not an inspiring one if, in the outcome, the scientist is routed. That is why some of the most important biological problems have not yet appeared on the agenda of practical research."

Having myself been pretty comprehensively routed in attempts to relate the organization and development of complex behavior to underlying genetic architecture, I feel I understand why functional approaches are so popular. After all, the growth of modern behavioral ecology is a most natural extension of classical ethology. (I should declare here that I don't wish to distinguish between "behavioral ecology," "sociobiology," and "evolutionary psychology." We may note that there are now at least *three* specialist journals and that many of the papers in *Animal Behavior* concern behavioral ecology.) It often involves field work, it is always concerned with the way an animal's behavior has become adapted to maximize its fitness. There is much sophisticated modeling by new generations of mathematically more apt ethologists, many of them derived from Maynard Smith's (1982) invaluable development of the concept of the Evolutionarily Stable Strategy. I believe Tinbergen would have loved this elegant and precise way of linking function and evolution in exactly the way he demanded.

Modern studies have brought the ethological approach into areas of social behavior and the evolution of social systems which were scarcely touched at the time Tinbergen was writing. His own book entitled *Social Behaviour in Animals* (1953) was almost totally confined to dyadic encounters between parents and offspring, sexual partners, or rivals. Even at the level of the individual, the idea of function is now extended far, far beyond the way we once imagined it. Thus, classical ethology would interpret the feeding behavior of a bird in terms of appetitive searching behavior interrupted at intervals by its performing rather stereotyped feeding movement in response to the external stimulus of a food item it has perceived. This remains largely true, but it is only a fraction of the picture. The development and modeling of optimal foraging theory reveals that a bird may constantly be making second-to-second decisions about its feeding. How long should it stay in one place searching and when should it give up and move to another place? The answers will depend amongst other things on how hungry it is, its past experience, the distribution of the food items and their quality. What is remarkable is to discover that, when such behavior is modeled and predictions made as to the best strategy given certain parameters, the details of the bird's behavior often match them extremely well.

This is but one example of the sophisticated approach to function which now concerns many ethologists. It relies on the old ethological skills of observation and careful description, often combining field work and laboratory experiments, but goes far beyond to the examination of long sequences of behavior, the allocation of priorities and the trade offs between them leading to different outcomes. None of this would, I feel, surprise Tinbergen who often took delight in showing just how perfectly behavior was matched to function. Black-headed gulls spend approximately 5 minutes per year removing eggshells from their nest, but he showed in detail how this behavior and its exact timing after the chicks hatched was adaptive against predators (Tinbergen *et al.*, 1962).

The allocation of function on such a broad scale and to such subtleties of behavioral changes even from second to second has not been without its critics. Gould and Lewontin (1979), using an interesting architectural metaphor – the Spandrels of San Marco – suggested that, with the eye of faith, it was all too easy to suggest functions for behavior, which might just be inevitable results of situation and conditions. They attacked some of the wilder shores of sociobiology on this account. The association of some behavior pattern or some structure with a particular behavioral outcome must not lightly be assumed to be an evolved, functional link. This is particularly the case when there are several possible functions and it may be well nigh impossible to distinguish between them by observation or experiment and in any case more than one function may be involved. One thinks of the behavior of helpers at the nest, for example (see Jamieson and Craig, 1987). A necessary caution then, but this should not inhibit the construction of functional hypotheses. "Everything is likely to be adaptive," is not a bad slogan to start out with so long as you keep a cool head! After all, the history of biology is full of examples where function was initially dismissed only to be amply proven by proper observation and experiment. Thus it had been suggested that insects had no color vision but von Frisch could not believe that the colors of flowers were without function. An early description of the beak of a crossbill (*Loxia curvirostra*) referred to it as "an abomination of nature" and so on. Robinson (1991) gives a number of examples of puzzling structures whose function is unfathomable until one watches behavior and sees them in action, just the course that Tinbergen recommends.

Modern behavioral ecology has established an astonishing body of data on the function and evolution of behavior; as we have described there has been no shortage of progress on these two of Tinbergen's questions. Even from the earliest days, ethology's emphasis on instinctive behavior presented formidable problems for the other two, mechanism (to be roughly equated with the causation question here) and development. It is certainly the case that there are now many beautiful studies showing how complex patterns of behavior develop. The crucial timing of various experiences for the development of bird song (Catchpole and Slater, 1995), the nature of sensitive periods and experience in sexual imprinting or the development of social behavior (see Bateson and Martin, 2002), and so on. At every stage we can see how the animal's genetic potential interacts with its physical and social environment to shape its behavior. Such studies exemplify some of the very best of modern ethology. One set of problems which remains concerns those elements of instinctive behavior which are typically highly species-specific and whose variations in form cannot be accounted for by the environment or experience during development. Lorenz was one of the first to emphasize that behavior, like morphology is a "property" of a species and evolves like it. Just as different duck species reared similarly will develop their characteristic plumage features, so will they develop the characteristic courtship behavior patterns. Like the plumage, these are clearly similar but they have striking and consistent species-specific features present from the outset when the birds are mature.

How can behavior appear in appropriate form and to the appropriate stimulus situation, fully fledged, as it were, at its first performance? Obviously, there must be a considerable genetic component to its development. How can genes coding for proteins, code for behavior? In the mid 1960s the distinguished geneticist I.M. Lerner entitled an address to the Behavior Genetics Association, "Two cheers for behavior genetics." I think this is still a just portion: as far as the direct question given above is concerned, the field has not had a very productive history. There is much sophisticated genetics which through selection and hybridization enables us to apportion a behavioral character's variance into genetic and environmental components and to explore the nature of its genetic architecture. The variation which is being analyzed is entirely quantitative and we can often identify "quantitative trait loci" of greater or lesser effect. For almost any behavior we find that numerous loci affect its level of expression. There have been many studies in which fast-breeding and easily maintained animals – Drosophila and mice especially – have been subjected to artificial selective breeding for behavioral traits. Most have succeeded, sometimes dramatically, in altering the frequency of performance of instinctive behavior, although the behavior itself stays obstinately intact. Genetic changes seem commonly to alter the threshold for the elicitation of behavior patterns but when they appear their form is unchanged. Bakker's (1986) heroic genetic analyses of stickleback aggressive behavior provide a good example. By selective breeding in different contexts he was able to show that the same genes sometimes overlap to affect aggression in more than one situation, sometimes more uniquely. This is a valuable result because it indicates how behavior's evolution may be constrained but it remains impossible to say much about the inheritance of the aggressive behavior patterns themselves.

The direct question asks how an animal is programmed to produce a repertoire of very stereotyped motor patterns in specific situations and often to a very specific set of stimuli – the classical ethological picture. Now clearly all kinds of interactions with the environment will be involved in the development of such patterns. Studying these will help us to identify more precisely the points at which genetically based information comes into operation. The diverse studies on bird song development offer many splendid examples and there are others in this volume. However, at the end we are often left with a very inaccessible problem of developmental neurobiology. How do genes so control the development of the nervous system that all the ducks or gulls or cichlid fish share some common repertoire of courtship movements but each species has emphasized elements of them in particular ways? We should not lightly dismiss the possibility that young individuals can copy from others (there are such instances in the behavior of birds and mammals) but here this is very improbable. It is one type of situation in which the Kaspar Hauser experiment can eliminate some routes of development. Genes must somehow "hard wire" the nervous system to operate in a very particular way. Expressed in this way, we are reiterating the exact kind of point which was made by Lorenz and others 40 years ago and we are little nearer to any kind of answer. Sometimes it has been possible to hybridize species whose instinctive repertoire is distinct at least in part. The songs of some crickets and of some pigeons are examples. Sometimes we see fairly clear dominance of one parent's behavior, more often we observe that a mixture, perhaps one should say jumble of genes in the F1 leads to a similar breakdown of behavior. It is not very informative and rarely is it possible to move on to an F2 generation where independent segregation from the two parental genotypes might suggest behavioral "units" of some kind.

The spectacular rise of molecular genetics has lent a new impetus to studies of single gene mutations and their effects on behavior. Now it is often possible to describe the gene's products and something of its actions on the nervous system, even the site of its action on occasions. This modern stage of the field is well summarized by Sokolowski (2001). However, it remains the case that, as outlined above, for the most part the genes described modulate the expression of behavior only and tell us very little about how the nervous system is organized to produce it. The exception would be mutant studies on learning, where some genetic studies are indeed giving insights into neural mechanisms, but learning is a more general process function of the nervous system than are the specifics of instinctive patterns. Nevertheless, it looks as if the genetic control of behavioral development might begin to reveal striking parallels with morphological development. Genes coding for the same polypeptides may be involved in the regulation of behavior across groups even across phyla, e.g., Drosophila, bees, fish and rodents; see Robinson and Ben-Shahar (2002). Perhaps we must consider the possibility that there are behavioral homeoboxes!

I suspect that the mechanisms by which genes "build" a nervous system that can mediate instinctive behavior will remain elusive for a long time to come. They represent an example of Medawar's class of intractable problems. However, I suppose we can at least *in principle* imagine how genetic programming could wire up a developing nervous center to produce a specific type of stereotyped motor pattern. For an examination of the possibilities it is certainly worth revisiting Hoyle's (1984) lively setting out of the neurophysiological background to stereotyped species-specific behavior patterns and the discussion it provoked. Such a developmental phenomenon might eventually become accessible to the techniques of developmental neurobiology. But some modern studies of behavioral ecology require us to demand far more of the genes. It is quite certain that animals sometimes make a subtle choice of their response depending not just on the immediate situation, but on the circumstances which led up to it. Davies's (1992) beautiful studies of dunnocks (*Prunella modularis*) have shown that many females have two males on their territory. The alpha male is shadowed by a beta male who usually fathers some of the offspring. The beta male matches the amount of help he gives when feeding young to the amount of contact – "mating access" – he had during her most fertile period during egg-laying! Functionally, this obviously makes sense, but surely it raises formidable developmental questions.

Dunnocks are short-lived in the wild; there is no possibility of a male trying out various feeding strategies and learning the best outcome. Somehow he must inherit the ability to match "mating access" to helping at the nest several days later. This is just one well-worked example but there are numerous instances of short-lived animals making such choices. The optimum feeding strategies mentioned above provide many examples. The observations lead to clear hypotheses about behavior which work out in practice and the function and evolution of the behavior are well established, but what of its development and causation? In such cases we tend to think of the simplest way that such behavior could be organized. We do not invoke conscious choice but fall back upon "rules of thumb." For example, Richard Dawkins once suggested that what genes might code for in the parental behavior of passerine birds is, "feed conspicuous gapes inside the nest rim." It is less easy to come up with a rule for the dunnocks but Davies's own studies have begun to close in on what aspects of the situation beta males might be responding to. However daunting the developmental problems, good ethological observation and experiment will remain one important way forward.

In conclusion, I turn to the other neglected aspect of Tinbergen's causation question – motivation, by which I mean to include "drives" in the old ethological sense. The rather simple exposition of instinctive drives set out by Lorenz and Tinbergen quickly proved to have severe limitations. Hinde's (1959, 1960) key papers criticizing the concepts were very influential and indeed some regarded them as having administered the *coup de grâce*. Slowly at first, but then more rapidly in the early 1970s neglect set in. Hogan (2005, this volume) describes this well. However inadequate were the original formulations, the phenomena they attempted to deal with remain as vivid as ever. In the first edition of her book which covered the modern approach to ethology, Marian Dawkins (1986) discussed them in a chapter she entitled "Some obstinate remnants." That rather makes the point; there is something to explain which won't go away! Thus, the old conflict hypothesis for the origins and causation of threat and courtship displays which postulates the simultaneous arousal and conflict between systems controlling aggressive and fearful responses retains considerable explanatory power for some situations. See, for example, Baerends (1985) on vertebrates and Maynard Smith and Reichart's (1984) study of spiders. Lorenz's concept of "action specific energy" building up over time and thresholds of response falling accordingly was attractive to behavior workers in the early days of ethology precisely because it seemed to model what they observed in nature. Deprived of suitable nest material a caged finch or ring dove will attempt to carry scraps of food, even its own feces to the empty nest site and perform the movements of building there. Now sometimes analysis of such situations has shown that what "switches off" the behavior is not its performance (as Lorenz's model suggested) but the animal's detection that a result has been achieved, i.e., in the example given here, it might be the presence of a nest at the site. Nevertheless, in the absence of such "consummatory stimuli" some central process must result in the compulsion – it is hard to regard it otherwise – to nest build. Workers in what is now called "applied ethology" and concerned with the welfare of domestic animals have to face up to such issues all the time. Intensively reared chickens present many examples. Are hens distressed if they are unable to lay their eggs in a nest box, or dustbathe, or scratch for food? Certainly answers to questions of this type are not going to come from much of modern ethology. Domestic animals, removed from almost any trace of the environment in which their ancestors evolved, retain elements of what was adaptive behavior, and the associated motivational states. Applied ethologists try to put animals into situations where they can reveal their motivation by their choices, thus finding ways which enable them to express themselves and thereby improve their welfare.

The trouble is that the concept of an animal's "welfare" almost inevitably leads us to contemplate that, apart from its physical health and well-being, it may share with us those feelings which go with frustration and confinement. For us, powerful motivation is associated with emotion, emotion which has easily observable physiological consequences. Similar consequences are easily observed in animals too which raises all kinds of questions about their subjective experiences. By far from logical links I find myself thinking over a whole new body of modern research which concerns itself with complex cognitive capacities of animals and all the mental processes which might go with them. There is now much speculation and experiment directed towards the possibility that some animals have a mind with consciousness of their own existence and that of others outside themselves.

For ethologists the modern approach to these issues began with Don Griffin's (1976) book *The Question of Animal Awareness* (which may have been a deliberately provocative title) and a subsequent review (Griffin, 1978). These caused a considerable and rather uneasy stir amongst ethologists at the time. Of course Griffin was addressing issues which go much further back; the higher cognitive powers of animals and their subjective feelings were commonly assumed during the 19th century. However there was an almost complete rejection for most of the 20th especially amongst experimental psychologists for whom any trace of subjective thinking in regard to animals was anathema. Nevertheless Griffin's bold return to face issues which never go away struck a chord with some ethologists. It certainly emboldened them to undertake research which attempts to explore whether some animals have higher mental capacities. They see this approach as a direct extension of ethology itself, hence the title of Griffin's (1978) review and the Festschrift Ristau (1991) edited for him which is called *Cognitive Ethology: The Minds of Other Animals*. This now comprises a very large literature and the best of it does strive to retain objectivity. Its proponents approach their research with Tinbergen's four questions firmly in mind. Unsurprisingly, the cluster of topics that comprise cognitive ethology attract much attention from those researching the welfare of domestic animals. Marian Dawkins (2006) provides a most valuable review which goes beyond just the welfare issues. The most difficult and the most contentious question remains that of animal consciousness. and, almost needless to say, the present position includes very extreme views on both sides.

I include cognitive ethology here because I cannot help wondering what Tinbergen himself would make of it, especially the debates around animal consciousness. Kruuk (2003) believes he would certainly *not* have approved and I agree with him absolutely. At the outset of *The Study of Instinct* (1951) Tinbergen dismisses any approach to subjective phenomena in animals, suggesting that, as it is impossible to observe them objectively, "it is idle either to claim or to deny their existence." Twelve years later in the Four Questions paper he sticks rigorously to his original definition of ethology as the objective study of the biology of behavior.

So perhaps there is one area of ethology which is going on into pastures new. No matter, the original four remain unchallenged. Although Medawar's rule applies and not all of them are equally easily accessible, they are nonetheless all equally important. This conclusion is developed and abundantly illustrated in the essays which make up this volume. The Maestro certainly *would* have approved for they show that his injunction retains its relevance and a fully rounded ethological approach remains as valuable as ever.

Aubrey Manning

# Preface

This volume brings together a collection of papers written by contributors to a symposium entitled "Evolution, function, development, causation: Tinbergen's four questions and contemporary animal biology," held at the Institute of Biology, Leiden University, the Netherlands on 5 September, 2003. The symposium was organized by The Royal Dutch Zoological Society (KNDV) with the Dutch Society for Behavioural Biology (NVG), to commemorate the 40th anniversary of the publication of Niko Tinbergen's seminal paper "*On aims and methods of ethology.*" Leiden was a fitting venue for the symposium, because Tinbergen held a chair there before he left in 1949 to become a demonstrator in the Department of Zoology at Oxford. Moreover, the symposium was held at the "van der Klaauw laboratory," and it was Professor Cornelis van der Klaauw who invited Konrad Lorenz to a symposium on "instinct" held in Leiden in November, 1936. That was the first time that Tinbergen and Lorenz met, a meeting that culminated in a life-long friendship. Indeed, Tinbergen's "aims and methods" paper was dedicated to Lorenz on the occasion of his 60th birthday.

In his paper, Tinbergen discussed the field of ethology, now usually known as behavioral biology, and defined it as "the biological study of behaviour." The "aims and methods" paper is best known for the identification of four major problems in the study of behavior: causation, development (ontogeny), function (survival value), and evolution. The main body of the paper comprises his views on each of these problems and, 40 years hence, is still a joy to read and has lost none of its brilliance. Tinbergen tackles difficult and sometimes controversial issues with wit and humor. In doing so, he set the agenda for animal behavior research that is still very much relevant in the 21st century. In fact, we would maintain that every student of animal behavior should attempt to grasp the message of this ground-breaking paper before embarking on a research project. Tinbergen's aim with this paper was to "attempt an evaluation of the present scope of our science." He considered such an attempt worthwhile because "ethologists differ widely in their opinion of what their science is about." Indeed, in a different sense this still holds true today, in that many scientists studying topics closely related to the four major problems of behavioral biology probably do not consider themselves ethologists, but instead see themselves as, e.g., neurobiologists or ecologists. Indeed, Tinbergen himself already recognized the overlap between research fields, stating that "I have used the word 'Ethology' for a vast complex of sciences (...), such as certain branches of Psychology and Physiology." Perhaps one of the greatest challenges behavioral biology faces today is the integration of such widely diverging fields while maintaining the identity of "Behavioral Biology" as a research field.

Appropriately, the book opens with a facsimile of the classic paper by the "Maestro" – as Tinbergen was fondly known by his Oxford students; see Hans Kruuk's excellent biography *Niko's Nature* (2003). After a general introductory chapter by Hogan and Bolhuis, the following five chapters are concerned with Tinbergen's four questions. Hogan, Crews and Groothuis, Hogan and Bolhuis, Cuthill, and Ryan address Causation, Development, Function, and Evolution, respectively. These chapters clearly show how Tinbergen's ideas have inspired these authors and influenced their own thinking and research. Tinbergen readily acknowledged his debt to Julian Huxley for proposing causation, evolution, and survival value as the three main problems in biology, with Tinbergen adding development. The addition of development to Huxley's list of three suggests that Tinbergen thought it was an important problem in animal behavior. We agree – hence there are two chapters concerned with behavioral development, with the article by Crews and Groothuis addressing the topic of sexual and individual differentiation and the one by Hogan and Bolhuis providing a more general overview of developmental research since Tinbergen's paper. The chapter by Sherry concerns work involving an integration of Tinbergen's four questions. Finally, Bolhuis discusses some key issues in the debate that ensued after his critique of the integrative approach, and he evaluates recent relevant evidence bearing on this debate.

The symposium – and hence this volume – would not have been possible without the help of many different people. We are grateful to the board of Leiden University, to Frans Saris, then Dean of the Faculty of Natural Sciences, to the Scientific Director of the Institute of Biology,

Eddy van der Meijden, and to the current leader of the behavioral biology group at Leiden, Carel ten Cate, for their support and for hosting this meeting in their Institute. Thanks are due to the Netherlands Organization for Scientific Research (NWO) and to the Royal Netherlands Academy of Sciences for their generous support for the meeting. Also, we would like to thank our co-organizers, the Dutch Society for Behavioural Biology, and its former President, Dr. Menno Kruk. In addition, we would like to thank our fellow members of the Board of the Royal Dutch Zoological Society for their hard work in organizing the symposium: Johan van Leeuwen, Joris Koene, Thijs Zandbergen and particularly Evert Meelis; without his hard work and organizational skills, the symposium would not have been possible.

We are most grateful to the authors for their contributions. In addition, we thank Aubrey Manning – a leading figure in behavioral biology and a student of Niko Tinbergen – for writing a wonderful Foreword. We are extremely pleased that we could reprint the original Tinbergen paper, for which we are very grateful to Blackwell Publishing.

Utrecht, J. J. B.
Groningen, S. V.

*Department of Zoology, University of Oxford*

# On aims and methods of Ethology

By N. Tinbergen[1])

*Received 16 March 1963*

Ethology, the term now widely in use in the English speaking world for the branch of science called in Germany „Vergleichende Verhaltensforschung" or "Tierpsychologie" is perhaps defined most easily in historical terms, *viz.* as the type of behaviour study which was given a strong impetus, and was made "respectable", by KONRAD LORENZ. LORENZ himself was greatly influenced by CHARLES OTIS WHITMAN and OSKAR HEINROTH — in fact, when LORENZ was asked at an international interdisciplinary conference in 1955 how he would define Ethology, he said: "The branch of research started by OSKAR HEINROTH" (1955, p. 77). Although it is only fair to point out that certain aspects of modern Ethology were already adumbrated in the work of men such as HUXLEY (1914, 1923) and VERWEY (1930), these historical statements are both correct as far as they go. However, they do not tell us much about the nature of Ethology. In this paper I wish to attempt an evaluation of the present scope of our science and, in addition, to try and formulate what exactly it is that makes us consider LORENZ "the father of modern Ethology". Such an attempt seems to me worthwhile for several reasons: there is no consistent "public image" of Ethology among outsiders; and worse: ethologists themselves differ widely in their opinions of what their science is about. I have heard Ethology characterised as the study of releasers, as the science of imprinting, as the science of innate behaviour; some say it is the activities of animal lovers; still others see it as the study of animals in their natural surroundings. It just is a fact that we are still very far from being a unified science, from having a clear conception of the aims of study, of the methods employed and of the relevance of the methods to the aims. Yet for the future development of Ethology it seems to me important to continue our attempts to clarify our thinking, particularly about the nature of the questions we are trying to answer. When in these pages I venture once more to bring this subject up for discussion, I do this in full awareness of the fact that our thinking is still in a state of flux and that many of my close colleagues may disagree with what I am going to say. However, I believe that, if we do not continue to give thought to the problem of our overall aims, our field will be in danger of either splitting up into seemingly unrelated sub-sciences, or of becoming an isolated "-ism". I also believe that I can honour KONRAD LORENZ in no better way than by continuing this kind of "soul-searching". I have not hesitated to give personal views even at the risk of being considered rash or provocative.

---

[1]) Dedicated to Professor KONRAD LORENZ at the occasion of his 60th birthday.

## Ethology a branch of Biology

In the course of thirty years devoted to ethological studies I have become increasingly convinced that the fairest characterisation of Ethology is *"the biological study of behaviour"*. By this I mean that the science is characterised by an observable phenomenon (behaviour, or movement), and by a type of approach, a method of study (the biological method). The first means that the starting point of our work has been and remains inductive, for which description of observable phenomena is required. The biological method is characterised by the general scientific method, and in addition by the kind of questions we ask, which are the same throughout Biology and some of which are peculiar to it. HUXLEY likes to speak of "the three major problems of Biology": that of *causation,* that of *survival* value, and that of *evolution* — to which I should like to add a fourth, that of *ontogeny.* There is, of course, overlap between the fields covered by these questions, yet I believe with HUXLEY that it is useful both to distinguish between them and to insist that a comprehensive, coherent science of Ethology has to give equal attention to each of them and to their integration. My thesis will be that the great contribution KONRAD LORENZ has made to Ethology, and thus to Biology and Psychology, is that he made us realise this close affinity between Ethology and the rest of Biology; that he has made us apply "biological thinking" to a phenomenon to which it had hitherto not been as consistently applied as was desirable. This is, of course, not to belittle LORENZ's concrete, factual contributions, which we all know are massive, but I submit that the significance of all his contributions is best characterised by saying that he made us look at behaviour through the eyes of biologists. I also submit that this is an achievement of tremendous importance and that, if anything deserves the much-abused name of "a major breakthrough", LORENZ's achievement does.

I shall devote the next pages to some remarks on each of these four problems as they apply to behaviour. If these remarks appear to some readers unsophisticated, I should like to remind them of the fact that Ethology is a science in its infancy, where even a little plain common sense can help.

## Observation and Description

One thing the early ethologists had in common was the wish to return to an inductive start, to observation and description of the enormous variety of animal behaviour repertoires and to the simple, though admittedly vague and general question: "Why do these animals behave as they do?" Ethologists were so intent on this return to observation and description because, being either field naturalists or zoo-men, they were personally acquainted with an overwhelming variety of puzzling behaviour patterns which were simply not mentioned in behaviour textbooks, let alone analysed or interpreted. They felt quite correctly that they were discovering an entire unexplored world. In a sense this "return to nature" was a reaction against a tendency prevalent at that time in Psychology to concentrate on a few phenomena observed in a handful of species which were kept in impoverished environments, to formulate theories claimed to be general, and to proceed deductively by testing these theories experimentally. It has been said that, in its haste to step into the twentieth century and to become a respectable science, Psychology skipped the preliminary descriptive stage that other natural sciences had gone through, and so was soon losing touch with the natural phenomena.

Ethology was also a reaction against current science in another sense: zoologists with an interest in the living animal, overfed with details of a type of Comparative Anatomy which became increasingly interested in mere homology and lost interest in function, went out to see for themselves what animals did with all the organs portrayed in anatomy handbooks and on blackboards, and seen, discoloured, pickled and "mummified" in standard dissections.

Much of the early ethological work contained a good deal of description and, in these first days of reconnaissance, of taking stock, we tended to think of "ethograms" as hundred-page papers which could contain about all we wanted to know about a species. Even this modest aim, a very sketchy description, was reached for very few species only. We must hope that the descriptive phase is not going to come to a premature ending. Already there are signs that we are moving into an analytical phase, in which the ratio between experimental analysis and description is rapidly increasing. This is a natural outcome of LORENZ's own work, and it is, of course, imperative that work on causation should be intensified and refined. However, we would deceive ourselves if we assumed that there is no longer a need for descriptive work. Misgivings about this wholesale swing towards analysis have been expressed, for instance by NIELSEN (1958), who wrote: "In 'modern' Ethology nobody pays the slightest attention to anything but the 'why'. It is a very peculiar situation: we have a science dealing with the causal explanation of observations but the collection of the basic observations is no longer considered a part of the science" (NIELSEN 1958, p. 564). While at first glance this is a surprising remark, which very few Ethologists and non-Ethologists will agree with, we cannot brush it aside entirely.

The issue is admittedly not a simple one. Description is never, can never be, random; it is in fact highly selective, and selection is made with reference to the problems, hypotheses and methods the investigator has in mind. In the early days of Ethology these limitations of our descriptions were not always obvious — mainly, I believe, because most of us were not sufficiently conscious of our limited aims, and certainly were not sufficiently aware of the criteria we used for selection.

The variety of behaviours found in the animal kingdom is so vast, and their description is so much more laborious than the description of structure, that selectiveness of description will become increasingly urgent. This will only be possible by a more explicit formulation of the problems we wish to study, and by growing certainty about the nature of the data we need. Yet even with the most economic procedures the amount of description to be done will long remain very large, so large in fact that we shall soon have to resort to a policy of filing descriptive material in libraries or archives (including film libraries) rather than publishing it in the usual journals. Already there are journals which demand a reduction of descriptive material to the absolute minimum required for an understanding of the experiments reported on (or even to less than this minimum); the descriptions (and often the argument behind the work) one has to pick up in personal conversations at conferences.

However, if we overdo this in itself justifiable tendency of making description subject to our analytical aims, we may well fall into the trap some branches of Psychology have fallen into, and fail to describe any behaviour that seems "trivial" to us; we might forget that naive, unsophisticated, or intuitively guided observation may open our eyes to new problems. Contempt for simple observation is a lethal trait in any science, and certainly in a science as young as ours.

It seems to me that one of the lessons we can draw from LORENZ's work is that our science will always need naturalists and observers as well as experimenters; we must, by a balanced development of our science, make sure that we attract the greatest possible variety of talent, and certainly not discourage the man with a gift for observation. Instead we should attract such men, for they are rare; we must encourage them to develop their gifts of observation and help them ask relevant questions with respect to what they have seen.

## Causation

At an early stage in his work (e.g. 1935, 1937) LORENZ made three statements which I should like to emphasise because I think that modern Ethology derived much of its inspiration from them: (1) animals can be said to "possess" behaviour characteristics just as they "possess" certain structural and physiological characteristics; while LORENZ emphasised this particularly for the relatively stereotyped motor patterns which he called „Instinkthandlungen" we know now that it applies to many other aspects of behaviour; (2) what we call behaviour is, even in its relatively simple forms, something vastly more complex than the types of movements which were then the usual objects of physiological study (and this applied equally to the sensory, the motor and the central nervous processes involved); and (3) the initiation, coordination and cessation of behaviour patterns are controlled by the external world to a lesser extent than reflex-physiologists were at that time prepared to admit.

I should like to elaborate these points to some extent.

(1) The first statement was based on an unrivalled store of first-hand experience as well as on much that had already been published, notably in the works of HEINROTH (1911) and of WHITMAN (1919). It led LORENZ to consider behaviour patterns (and by implication the mechanisms underlying them) as *organs*, as attributes with special functions to which they were intricately adapted. This again facilitated causal analysis without interference by either subjectivism or teleology. By subjectivism I mean here the procedure of replying to the question "What causes this behaviour?" by referring to a subjective experience, i.e. a process which per definition can be observed by no one except the subject. It seems to me worth pointing out that Ethology has not yet completely succeeded in freeing itself from subjectivism in this sense. It is true that one rarely meets with it in its crudest form ("the animal attacks because it feels angry"), but in its subtler forms it is still very much with us. Concepts such as "play" and "learning" have not yet been purged completely from their subjectivist, anthropomorphic undertones. Both terms have not yet been satisfactorily defined objectively, and this might well prove impossible; both may well lump phenomena on the one hand, and exclude other phenomena on the other hand (and thus confuse the issue by a false classification) simply because the concepts are directly derived from human experience. In both fields the growing tendency to ignore the term and to return to the phenomena (which are singled out for study because they are suspected of having a different causation than other phenomena), is, I think, an inevitable result of the consistent application of biological thinking to behaviour.

Teleology also can be said to have ceased to be a source of confusion in its cruder forms, in which function was given as a proximate cause, but it may well be a major stumbling block to causal analysis in its less obvious forms. Throughout Biology we tend to classify, and hence to give names, on the basis of criteria of common function. The more complex the behaviour systems we deal with, the more dangerous this can be. For instance, although the more sophisticated ethologist is fully aware of the fact that the term "Innate Releasing Mechanism" refers to a type of function, of achievement, found in many different animals; and that different animal types may well have convergently achieved "mood controlled selective responsiveness" by entirely different mechanisms, the term has given rise to misunderstanding on this point, as LEHRMAN's criticism (1953) showed. And who knows what different mechanisms we are lumping under terms such as learning, displacement activity, drive-reduction?

Another type of difficulty which may have to do with our thinking in terms of function is caused by our habit to coin terms for major functional units such as nest building, fighting or sexual behaviour and treat them as units of mechanism. For instance, while the fact that all fighting acts fluctuate together in the natural situation, as do all components of escape, *does* justify us to use "fighting tendency" and "escape tendency" when we are involved in the first step of analysis of movements caused by the *simultaneous arousal of the two*, as soon as we begin to analyse the causation of each of them separately, it is pre-judging the issue if we take for granted that each is in itself a causally closely-linked complex of components. We may then find that some components have closer links with processes outside this functional system than with other parts of the same functional system (see BEER [1962] on incubation and HINDE [1958, 1959] with respect to nest building). Our habit of giving names to systems characterised by an achievement has made thinking along consistent analytical lines much more difficult than it would have been if we could have applied a more neutral terminology. But rather than to advocate such a dry, non-committal terminology, I would like to accept any frankly functional term, as long as this is done consciously. No physiologist applying the term "eye" to a vertebrate lens eye as well as to a compound Arthropod eye is in danger of assuming that the mechanisms of the two are the same; he just knows that the word "eye" characterises achievement, and no more.

The treatment of behaviour patterns as organs has not merely removed obstacles to analysis, it has also positively facilitated causal analysis, for it led to the realisation that each animal is endowed with a strictly limited, albeit hugely complex, behaviour machinery which (if stripped of variations due to differences in environment during ontogeny, and of immediate effects of a fluctuating environment) is surprisingly constant throughout a species or population. This awareness of the repeatability of behaviour has stimulated causal analysis of an ever-increasing number of properties discovered to be species-specific rather than endlessly variable.

It may not be superfluous to stress that the recognition of the existence of many species-specific behaviour characters does not necessarily imply that all these characters are "innate" in the sense of ontogenetically wholly independent of the environment. It is true that this is often assumed in many ethological publications, and I shall have to return to this when I discuss behaviour ontogeny, but this point is irrelevant to us here. LORENZ's emphasis on the fact that so much in behaviour mechanisms is species-specific remains as fruitful as ever.

(2) LORENZ's emphasis on the complexity of behaviour phenomena (which is only seemingly contradictory to his inclination towards simplifying physiological *explanations*), seems to me still to be of the greatest importance, even though we now take this complexity for granted. Lack of appreciation of this point seems to me to have been one of the most important reasons for

the lack of co-operation between physiologists and ethologists. The magnitude of the gap between the phenomena studied by ethologists and those studied by nenurophysiologistshas been underrated by both parties. The early etho-logists underrated the complexity of behaviour mechanisms in various ways, as was evident from our early attempts at "physiologising". One example of this is the lack of any provision for negative feedback in LORENZ's original "psychohydraulics" model (1950); another is provided by my own sketch (1951) of the organisation of the hierarchy in behaviour mechanisms; another again can be found in the original explanation of "displacement activities" (compare, for instance, TINBERGEN 1940, ROWELL 1961 and SEVENSTER 1961). Ethologists are now increasingly avoiding such over-simplifications, without however giving up the application of strictly analytical procedures which were started by LORENZ's work. A corollary of this is the development of concepts suited to the stage of analysis, concepts (and terms) which avoid implying physiological explanations — VON HOLST's (VON HOLST and v. ST. PAUL 1960) *"niveau-adäquate Terminologie"* — without becoming enslaved by such terminology and shutting the door to further analysis. Until recently neuro-physiologists, concerned with the analysis of relatively simple processes, were either not considering the more complex phenomena, or were too ready to assume that combinations of the basic phenomena they knew would some day be found to account for behaviour of the intact animal. A striking example of the latter attitude was quoted by VON HOLST and MITTELSTAEDT (1950) in their analysis of the way in which an optomotor response was found not to interfere with "spontaneous" locomotion. Until these authors checked by a simple experiment whether it was true that, under such circumstances, the optomotor response was inhibited (and found that this was not so) this unproven hypothesis seems to have been taken for granted.

The situation is now changing rapidly. The "no-man's land" between Ethology and Neurophysiology is being invaded from both sides. While etholo-gists are making progress with the "descending" breakdown of complex pheno-mena, neurophysiologists are "ascending", extending their research to pheno-mena of greater complexity than was usual 20 years ago. To what extent the latter development has been influenced by Ethology is difficult to say, for who can trace the origin of new fashions in a science? While I am convinced that it would have happened anyway, I am equally convinced that the growth of Ethology has speeded up the process. The *rapprochement* between the two fields has gone so far already that it begins to be difficult, and in some cases even impossible, to say where Ethology stops and Neurophysiology begins. Are DETHIER and ROEDER ethologists or neurophysiologists? And where to put VON HOLST, MITTELSTAEDT and HASSENSTEIN? Several of my colleagues are still inclined to draw a sharp line between the two fields, and to deny some of these workers a place among the physiologists, mainly, I understand, because the mechanisms they describe cannot yet be expressed in physico-chemical terms. I believe that this view is a denial of the fact that much which is conventionally called Physiology has not reached that stage either; or was accepted as part of Physiology before physico-chemical explanations were possible or even within sight. What happens throughout this entire field is that *achievements* of complex systems are, after a varying number of analytical steps, described in terms of *achievements* of component systems. If VON HOLST's work on the superposition effect (1937), HASSENSTEIN's work on the interaction of ommatidia responsible for the response to movement (1951, 1957), and VON HOLST's and MITTELSTAEDT's work on reafference (1950) is not Physiology, then why was SHERRINGTON's work Physiology? It is, of course, in itself completely unimportant whether or not one calls a certain type of work by a special name, as long as one agrees that it has a place in the progress of science, but the issue has important implications. I believe that it is doing our science a great deal of harm to impose boundaries between it and Physiology where there are none, or rather where there is only a "cline" from behaviour analysis on the one extreme to "Molecular Biology" on the other. I believe that the only criterion by which these extremes and the intermediate fields can be distinguished is that of the level of integration of the phenomena studied. For an understanding of our aims it seems to me much more important to recognise that fundamental identity of aims and method unites all these fields. It is the nature of the question asked that matters in this context, and this is the same throughout. Co-operation between all these workers is within reach, and the main obstacle seems to be lack of appreciation of the fact that there is a common aim.

(3) LORENZ's third postulate, stressing the part played in the control of behaviour by internal causal factors, has also had, and is still having, an effect on analytical studies. Again, the earlier confident statements about the nature of these internal determinants were, in some respects, premature and were at best over-simplifications. Thus, we are now far removed from the simple idea that the effect of hormones on behaviour is no more than a simple, direct stimulation of target tissues in the c.n.s., for it is clear that roundabout effects — e.g. those mediated by hormone-induced growth processes in the sensory periphery (BEACH and LEVINSON [1950], LEHRMAN [1955], and receptor-mediated feedback phenomena (LEHRMAN [1961], HINDE [1962]) xenter into the causation of hormone-controlled behaviour. Nor do we believe any more that a complex behaviour system serving one major function, such as nest building in birds, should necessarily be controlled by one single, compact "centre" in the c.n.s. Yet it is surprising to see how interest in "spontaneous" activity of nervous tissue, and in units controlling entire behaviour patterns has grown, and even how many of LORENZ's suggestions about central control of complex behaviour prove to have more than a core of truth in them. To take but one example, BLEST's work (1960) on the interaction between the tendency to fly and the antagonistic tendency to settle in *Automeris* moths, which by elimination of all the known or suspected alternatives was concluded to be due to direct interaction between parts of the c.n.s. itself, illustrates a trend which, I am sure, owes much LORENZ's approach.

These briefly mentioned samples do indicate, I believe, how analyses of behaviour mechanisms which were initiated in the earlier ethological studies are moving towards a fusion with the fields conventionally covered by Neurophysiology and Physiological Psychology. As far as the study of causation of behaviour is concerned the boundaries between these fields are disappearing, and we are moving fast towards one Physiology of Behaviour, ranging from behaviour of the individual and even of supra-individual societies all the way down to Molecular Biology. There ought to be one name for this field. This should not be Ethology, for on the one hand Ethology has a wider scope, since it is concerned with other problems as well; on the other hand, ethologists cannot claim the entire field of Behaviour Physiology as their domain, for they have traditionally worked on the higher levels of integration, in fact almost entirely on the intact animal. The only acceptable name for this part of the Biology of Behaviour would be *"Physiology of Behaviour"*, and this name should be understood to include the study of causation of animal movement with respect to all levels of integration.

## Survival value

Lorenz's thesis that behaviour patterns, i.e. their mechanisms, ought to be considered "organs", and to be studied as such, has also had a beneficial effect on the study of the survival value of behaviour. In the post-Darwinian era, a reaction against uncritical acceptance of the selection theory set in, which reached its climax in the great days of Comparative Anatomy, but which still affects many physiologically inclined biologists. It was a reaction against the habit of making uncritical guesses about the survival value, the function, of life processes and structures. This reaction, of course healthy in itself, did not (as one might expect) result in an attempt to improve methods of studying survival value; rather it deteriorated into lack of interest in the problem — one of the most deplorable things that can happen to a science. Worse, it even developed into an attitude of intolerance: even wondering about survival value was considered unscientific. I still remember how perplexed I was upon being told off firmly by one of my Zoology professors when I brought up the question of survival value after he had asked "Has anyone an idea why so many birds flock more densely when they are attacked by a bird of prey?"

Lorenz was never in danger of conforming to this fashion. He always was much too good a naturalist for this; he further had the good fortune of being taught by Hochstetter, an anatomist with a wide grasp of what Biology is about; finally, he was himself too clear a thinker to confuse teleology with the study of survival value. To him an organ was something which a species had evolved as one of its means for survival, something of which, as a matter of course, both the contribution it made to survival and its causation had to be studied. He has always been equally interested in "What is this good for?" and in "How does it work?". It was partly through this interest in survival value, for instance, that he arrived at his important concept of "Releaser" — an organ adapted to the function of sending out stimuli to which other individuals respond appropriately, i.e. in such a way that survival is promoted. "Releaser" was defined along much the same lines as any other effector, say, a wing, which is an organ adapted to the function of flying, or an endocrine gland, which is a gland adapted to the function of shedding hormones which, by acting on equally adapted "target organs", contribute to the proper co-ordination of functions within the body. It is an illustration of the inability of many biologists to think in terms of survival value that the concept of Releaser as an *organ characterised by a function* is so often misunderstood and confused with "anything which provides stimuli", a confusion due to the failure to see the function of the releaser and to preoccupation with the causation of the behaviour of the reacting animal.

It is through Lorenz's interest in survival value that he appealed so strongly to naturalists, to people who saw the whole animal in action in its natural surroundings, and who could not help seeing that every animal has to cope in numerous ways with a hostile, or at least unco-operative environment. Incidentally, just because Lorenz's work has revived interest in the study of survival value, and because this is an aspect of Ethology which may well fertilise other fields of Biology (where survival value studies are being neglected) I think it is regrettable that his fine new Institute has been named „Institut für Verhaltensphysiologie" — its field of research extends far beyond Physiology.

Being myself both a naturalist and an experimenter at heart, one of my primary interests has always been to find out, if possible by experimentation, how animal behaviour contributes to survival, and I shall therefore enlarge a little more on this theme than on the others.

I have always been amazed, and I must admit annoyed as well, when I met, among fellow-zoologists, with the implied or stated opinion that the study of survival value must necessarily be guesswork, and that exact experimentation on the problem is in principle not possible. I am convinced that this is due to a confusion of the study of natural selection with that of survival value. While I agree that the selection pressures which must be assumed to have moulded a species' past evolution can never be subjected to experimental proof, and must be traced indirectly, I think we have to keep emphasising that the survival value of the attributes of present-day species is just as much open to experimental inquiry as is the causation of behaviour or any other life process.

Our study always starts from an observable aspect of a life process — in the present case, behaviour. The study of causation is the study of preceding events which can be shown to contribute to the occurrence of the behaviour. In this study of cause-effect relationships the observable is the effect and the causes are sought. But life processes also have effects, and the student of survival value tries to find out whether any effect of the observed process contributes to survival if so how survival is promoted and whether it is promoted better by the observed process than by slightly different processes. It is clear that he too studies cause-effect relationships, but in his study the observable is the cause and he tries to trace effects. Both types of worker are therefore investigating cause-effect relationships, and the only difference is that the physiologist looks back in time, whereas the student of survival value, so-to-speak, looks "forward in time"; he follows events after the observable process has occurred. The crux is that both are concerned with a flow of events which can be observed repeatedly, and which thus, unlike the unique events of past evolution, can be subjected to observation and experiment as often as one wishes.

The fact that we tend to distinguish so sharply between the study of causes and the study of effects is due to what one could call an accident of human perception. We happen to observe behaviour more readily than survival, and that is why we start at what really is an arbitrary point in the flow of events. If we would agree to take survival as the starting point of our inquiry, our problem would just be that of causation; we would ask: "How does the animal — an unstable, 'improbable' system — manage to survive?" Both fields would fuse into one: the study of the causation of survival. Indeed, logically, survival should be the starting point of our studies. However, since we cannot ignore the fact that behaviour rather than survival is the thing we observe directly, we have, for practical reasons, to start there. But this being so, we have to study both causation and effects.

The widespread lack of interest in studies of survival value and the opinion that it can never move beyond the level of inspired guesswork are all the more puzzling because the literature contains quite a number of good experimental studies in which the survival value of behaviour has been as well demonstrated as anyone could wish. To mention just a few examples: MOSEBACH-PUKOWSKI (1937) has shown that the habit of crowding of *Vanessa* caterpillars has survival value in affording protection from insectivorous song birds: isolated caterpillars are eaten more readily than those living in a cluster. KRISTENSEN (TINBERGEN 1951) has shown that the "fanning" of male Sticklebacks has survival value by renewing the water round the eggs; if this is prevented the eggs die, and artificial ventilation saves them. BLEST (1957) has shown that sudden display of "eye spots" by certain moths scares away certain predators and thus saves lives. VON FRISCH's studies of the bee dance do not leave doubt about the survival value of this behaviour; it does direct workers to rich food sources unknown to them and thus greatly increases the efficiency of feeding; as LINDAUER (1961) has demonstrated, the dance also plays a part in directing a homeless swarm to a suitable site. Such studies are generally considered both interesting and reliable, and this gives the lie to the argument that survival value cannot be studied experimentally. What then are the reasons for this problem being underworked to such an extent?

First of all, the survival value of many attributes, behaviour and structure alike, is so obvious as to make experimental confirmation ludicrous. One need not starve an animal to death to show that its feeding behaviour has survival value, nor need one cut off a Blackbird's bill to show that this organ is necessary for successful feeding. But one of the reasons why ethologists are so much concerned with survival value is that the "use" of so many behaviour patterns is still completely unknown.

However, the quest for survival value involves, of course, much more than the demonstration that the Blackbird's bill is indispensable to it; one wants to know whether a bill of this size and this shape is best suited to feeding in the environment in which the Blackbird lives; similarly, one needs to understand in detail the suitability of every aspect of its feeding behaviour one sees, and this, of course, is very far from obvious. To think that we understand survival value completely in such cases is to think that, once it is obvious that sex hormones control mating behaviour, we need not inquire into the way they do this, nor into the interaction between various endocrine processes that are involved.

Another important reason for the lack of interest in survival value studies is a practical one. The method to demonstrate survival value of any attribute of an animal is to try whether or not the animal would be worse off if deprived of this attribute. This is easy with structures. For instance, BLEST could compare the effect of normal "eye spot"-bearing moths on song birds with that of moths whose eye spots were brushed off. In this test moths without eye spots could safely be regarded as differing from the controls in just this one respect. Similarly, HOOGLAND, MORRIS and TINBERGEN (1957) could show that Sticklebacks without spines were eaten more readily by small Pike than normal Sticklebacks. But how does one make an experimental animal which lacks just one behaviour pattern and is otherwise normal? How, for instance, to make a male bird which does not show aggressive behaviour, or lacks one of its threat postures, while in all other respects behaving normally?

This difficulty can in many cases be overcome, but one has to be aware of many pitfalls. Much of our evidence can come from systematic comparison of the success of animals at times when they do show a certain behaviour and the lack of success when they do not perform it. Thus, if a territory of a male bird is not invaded as long as it fights off intruders, but is invaded when, later in the day, his aggression wanes though he is still there and shows a variety of other behaviour, one has a good indication that it was the aggressive behaviour which kept the territory clear. Or, to take another example, if one can show that a motionless twig caterpillar is not eaten by birds while it is snapped up as soon as it moves (DE RUITER 1952) one can be pretty confident that immobility in this species has survival value.

Yet it is, of course, true that in such "natural experiments" one does not control the feature studied — one never knows which unknown aspect of the animal may have varied with the character studied; the aggressive bird may have made an ultrasonic sound; the moving caterpillar may have given off a scent. Therefore the method of studying the survival value of behaviour is to use dummies and to control their "behaviour". For instance, when a dummy of a male stickleback is either ignored, or merely approached, by a ripe female when moved at random, but elicits following and even the movements of creeping into the nest whenever it is made to move like a "nest-showing" male, even in the absence of any nest, then one has demonstrated the effect of nest-showing and, since eggs not laid in the nest are eaten or abandoned, one has shown that the male's nest-showing contributes to, and is even indispensable for successful reproduction. Of course, an experiment such as this is but the first step, for one needs to know the full story of the cause-effect relationship which makes the female respond the way she does; also, one wants to know not merely whether absence of the behaviour studied has an adverse effect; one also needs to know what kind of deviations from the natural behaviour would reduce the effect — which includes the task of finding out whether the natural movement has the optimal effect and, if not (as is the case in supernormal stimuli) why the behaviour is not "better" — a question which will crop up again in evolutionary studies.

So far, we have only made the barest beginning with this task; there are even many behaviour patterns of which we do not even know the basic answer: has it any function at all? As an illustration let me mention the example of the "rocking" of certain cryptic animals. There are a number of animals which, either as an introduction to the change from motionlessness to movement, or from movement to immobility, perform a series of curious rocking movements. HEINROTH (1909) described these for the Nightjar, and mentioned that they are also found in *Phyllium* and in *Dixippus*. BLEST (1960) has shown that many Saturniid moths have a similar movement, usually preceding settling. Now these animals are all camouflaged; many of their behaviour characters (immobility by day, background selection, semi-closing of the eyes in the Nightjar, etc.) are obviously adapted to the function of avoiding detection by visually hunting predators. In view of this the habit of rocking seems very strange indeed, for movement in general is a stimulus to which visually hunting predators react, and which these cryptic animals are for the  rest at such pains to avoid giving. Therefore the fact that these movements occur in such different animals suggests that they have survival value and are somehow connected with camouflage. I believe that a testable hypothesis can be formulated. DE RUITER (1952) has shown that European Jays ignore twig-shaped caterpillars as long as the latter stay motionless. However, this was only true of Jays which had grown up in a normal environment, and in particular those which had had the opportunity of discovering, by trial and error, that real twigs are inedible. Hand-raised Jays which had not "played about" with twigs took up and tried out twigs and caterpillars alike. We know that many young birds have, at the start, a very "open mind" with regard to food; they respond to an enormous variety of objects, edible and inedible alike, and learn to confine themselves to those they find edible. My suggestion is that we have as yet no more than the faintest idea of the kinds of things such birds learn when young. HEINROTH (1909) already suggested that the rocking movements of *Phyllium* might well be harmless because predators might recognise them as passive movements of a leaf slightly moved by the wind. It seems to me quite possible that many young birds actually do learn that certain types of movements are passive, and not indicative of animal prey, and that it might not only be harmless for certain cryptic animals to perform these movements but that it might be definitely beneficial to them because it might ensure that a predator sees them and concludes that they are just vegetable matter. A motionless cryptic animal "hopes to be overlooked"; a rocking cryptic animal makes sure that it is seen and ignored — which means survival.

Critical, scientific zoologists had a way of applying the term "armchair science" to such ideas; yet it is becoming increasingly clear that it was the critics who judged such issues without investigation and even without knowledge of the real events; it is obvious that a well planned study of the ontogeny of the feeding behaviour of certain predators, combined with an accurate study of these rocking movements and the passive movements of inanimate objects could prove the hypothesis to be right or wrong.

Of course, in selecting this example I have applied "shock tactics" by taking a very exceptional type of behaviour. However, I would like to submit, first that we know so little about behaviour that new, equally strange behaviour patterns could well be discovered in large numbers; second, that the problem of survival value applies equally to every detail of behaviour and structure, however self-evident or insignificant it might seem at first glance. For instance, the fact that the Godwit walks differently from the Lapwing would seem too trivial to pay attention to, yet KLOMP (1954) showed that these differences are adaptive: Godwits lift and fold their feet much more than do Lapwings and thus avoid getting their toes caught in the tall grass in which they breed. Lapwings avoid habitats with tall vegetation. A parent Kittiwake does not produce a sound when it is about to feed its young but other gulls do; the Kittiwake is the only species of gull whose chicks stay on the nest and therefore have not got to be called to food (E. CULLEN 1957). HEINROTH (1928) suggested that Starlings and Partridges show remarkably little inhibition from trying to walk or fly straight through the thin metal bars of bird cages because they are adapted to living among grass rather than trees, and grass gives way. The writings of the good naturalists are teeming with such hints, arguments, and occasional demonstrations of the functions of a multitude of aspects of species-specific behaviour. It is also the experience of every good naturalist that the longer one studies a species, the more adaptive aspects of its behaviour one becomes aware of. The phenomena are countless, the field is practically unexplored, and yet without exploring it systematically we cannot hope to understand how behaviour helps animals to survive.

How can we tackle this immense task and catch up with the backlog? Hypotheses can be arrived at in various ways. LORENZ himself has pointed out one which he derives from his particularly fruitful method of raising and keeping animals in freedom, yet in partly artificial surroundings. Under such conditions one observes a number of behaviour patterns which "misfire". The observer's reaction is: "This seems ill-adapted; but it must be good for something" and this makes him try to see the behaviour in its proper context.

The naturalist who studies animals in their natural surroundings must resort to other methods. His main source of inspiration is comparison. Through comparison he notices both similarities between species and differences between them. Either of these can be due to one of two sources. *Similarity* can be due to affinity, to common descent; or it can be due to convergent evolution. It is the convergences which call his attention to functional problems. This method has been applied beautifully by VON HAARTMAN (1957) in his study of adaptations in birds hole nesting. The *differences* between species can be due to lack of affinity, or they can be found in closely related species. The student of survival value concentrates on the latter differences, because they must be due to recent adaptive radiation. An example of this procedure is E. Cullen's elegant study (1957) of the peculiarities of Kittiwakes as compared with other gulls.

Such hypotheses can be made highly probable even without experimentation. E. Cullen's report contains hardly any experimentation, yet her conclusion that, for instance, immobility of the young, their "facing-away" gesture, the tameness of the adult Kittiwakes while on their cliff, their mud-trampling before nest building, and the absence of an acquired attachment to their own young are all adaptive corollaries of cliff-breeding carries conviction because this interpretation is the only one which fits into our general picture.

However, such studies are no more than a beginning; they can be extended and intensified in several ways.

First, as one becomes better acquainted with a species, one notices more and more aspects with a possible survival value. It took me ten years of observation to realise that the removal of the empty eggshell after hatching, which I had known all along the Black-headed Gulls to do, might have a definite function, and that even the length of delay of the response, which varies with the circumstances, and which is on the average longer than that found in the Ringed Plover, may be adaptive (Tinbergen et al. 1962).

Secondly, hunches about survival value must, where possible, be strengthened by experiment. This meets with obstacles of a practical nature, but once the need is obvious, ways can often be found. Thus when one sees that the breeding season of the Black-headed Gulls is much more synchronised than that of Gannets, and one notices at the same time that the late broods of Black-headed Gulls seem to be less successful than the majority; when one further has indications that such late broods perish through heavy predation, one can first systematically compare predation of late and "peak" broods, and ultimately design an experiment to find out whether or not synchronisation has survival value as an anti-predator defence.

Thirdly, the experimental demonstration of survival value involves quite a number of steps. Much of the experimental evidence is not complete, because it has (often of necessity) been done in a situation which differs essentially from the natural context. In order to study the survival value of egg-shell removal in the Black-headed Gull, my co-workers and I demonstrated that gulls' eggs, laid out well scattered over the hunting area of Carrion Crows and Herring Gulls, were found more readily when they hand an empty egg-shell at 4 inches distance than when no egg-shell was added (Tinbergen et al. 1962). However, before we can conclude from this that egg-shell removal reduces predation, we have to consider whether in the natural situation this is its only effect. When a gull removes the egg-shell it leaves its brood unguarded for a few seconds. This, we know, can be critical: neighbouring gulls or Crows sometimes snatch up an egg or a newly-hatched chick in a second or so. It clearly depends on the balance of advantage and disadvantage whether the response is on the whole useful or not. The strict test for this would be a comparison of breeding success of a population of gulls which remove the shell with that of a population which do not, though in all other respects identical to the shell-carrying population. It is just because this is impossible that we have to be content with less good evidence. There are several indications that the advantage outweighs the disadvantage. For instance, whenever a Crow is in sight, the tendency of the gulls to attack it and drive it off dominates that of removing the shell; also, the danger caused by leaving the nest for a few seconds might well be less than that caused by the presence of the tell-tale shells for a long time. It remains true, however, that the ultimate test of survival value is survival itself, survival in the natural environment. This ultimate test has been carried out in very few cases only; a good example, involving a colour adaption rather than a behavioural one, is supplied by the work of KETTLEWELL (1955, 1956) on differential survival of white and black mutants of *Biston betularia* in two different environments: one which favoured the white form, and one which favoured the dark form.

I have argued that survival value has to be studied in its own right, but there are two additional reasons. First, Zoophysiology derived, and again derives much of its inspiration and guidance from knowledge or hunches about survival value. Experiments on the external control of respiration often concentrate on the effect of varying oxygen and $CO_2$-content of the medium or in the blood; this is because one starts from the knowledge that it is oxygen the animal requires, and $CO_2$ it must get rid of. The work of VON FRISCH and his school on colour vision in the Honey Bee was set off by VON FRISCH refusing to believe that the colours of flowers had no function; our knowledge of the ability of Arthropods to register the plane of polarisation of light is due to VON FRISCH wondering about the exact function of the bees' dance.

Secondly, the part played by natural selection in evolution cannot be assessed without proper study of survival value. If we assume that differential mortality in a population is due to natural selection discriminating against the less well-equipped (the less "fit") forms, we have to know how to judge fitness, and that can only be done through studies of survival value.

To those, however, who argue that the only function of studies of survival value is to strengthen the theory of natural selection I should like to say: even if the present-day animals were created the way they are now, the fact that they manage to survive would pose the problem of how they do this.

### Ontogeny

A newly-hatched Herring Gull pecks selectively at red objects (GOETHE 1937, TINBERGEN & PERDECK 1950) but a human being has to learn to stop when the traffic lights turn red. We have to learn the intricately co-ordinated motor patterns of speech, whereas a Whitethroat raised in isolation produces the complicated normal song of its species (SAUER 1954). It is the contrast between man and animals in the ways they acquire either "knowledge" or "skill" which arouses in most of us an interest in the ontogeny of behaviour. As we all know, the systematic study of behaviour ontogeny has had a slow start, and for a long time was heavily weighted, but differently so in different groups of researchers. While animal psychologists explored the ways in which various types of learning might account for behaviour ontogeny, ethologists emphasised the unlearnt character of many aspects of animal behaviour. Ontogeny was, for a long time, and to a certain extent still is, a field in which there is a real clash of opinion. All concerned agree that a complete understanding of behaviour requires an understanding of its ontogeny, just as morphologists agree that it is not sufficient to understand the adult form, but also the way in which this develops during ontogeny. But there is no agreement about the nature of the problems involved, and while the methods applied by psychologists and ethologists begin to resemble each other so closely as (in some instances) to be indistinguishable, the interpretation of the results gives rise to much discussion.

I believe that this discussion has been and is still being bedevilled by semantics, and that it would be helpful if, instead of discussing the justification of the use of words such as "innate" and "acquired", of "instinct" (and "instinctive") and "learning" we could return to a statement of the phenomena to be understood and the questions to be asked — indeed I think this is imperative.

I should like to characterise the *phenomenon* as "change of behaviour machinery during development". This is not, of course, the same as a change of behaviour during development; when in spring we see a thrush pick up and smash a snail for the first time in months, this change in feeding behaviour may be due to snails having reappeared for the first time after winter. We can conclude that the thrush itself, i.e. its behaviour machinery, has changed only if the behaviour change occurred while the environment was kept constant. It should be pointed out in passing that systematic descriptions of behaviour ontogeny are still rare and fragmentary.

When we turn from description to causal analysis, and ask in what way the observed change in behaviour machinery has been brought about, the natural first step to take is to try and distinguish between environmental influences and those within the animal. It is about this very first, preliminary step that confusion has arisen.

As in studies of the cyclical behaviour of adult animals, external influences have been studied most, for the simple reason that they are so much more easily manipulated. It is also important to realise that, in ontogeny, the conclusion that a certain change is internally controlled (is "innate") is reached *by elimination*. This is not, of course, a reflection on the validity of a classification — one can perfectly well dichotomise any group of phenomena into one group possessing the character A and the rest not possessing A — but it does reflect on the justification of lumping all examples for which not all environmental influences have been eliminated into a class called "innate", thus suggesting a positive statement where merely a negative statement would be in order. And I submit that most statements about "innateness" of behaviour are based on the elimination of one or some out of several, perhaps many, possible external influences. I am again criticising myself just as much as others, for I am now convinced that I have helped to perpetuate the confusion. If we raise male Sticklebacks in isolation from fellow members of its own species, subject them as adults to tests with dummies, and find (E. CULLEN 1961) that they attack red dummies just as selectively as do normal males, we are entitled to say that exposure to red males cannot be responsible for the development of this selectiveness of response. We cannot, however, say anything about the problem whether or not interaction with the environment during "practising" has influenced the form of their fighting movements. When GROHMANN (1939) showed that the incomplete flying movements which young pigeons make while growing up (and which might be interpreted as providing "practice") did not influence their flying skill (birds that were prevented from flapping on the nest flew as well as controls on their first attempt), he eliminated a different form of interaction with the environment than CULLEN did with her Sticklebacks

It is not helpful and even wrong to apply to both behaviour patterns the term "innate", because in each case only one out of various environmental effects was excluded, and these were different in each case. The conclusion can only be formulated correctly in negative terms, in describing which environmental aspect was shown *not* to be influential.

There is, in addition, another reason for not applying the term "innate" to the fighting behaviour of the Kaspar Hauser Sticklebacks. KNOLL (1953) has shown that the rods of tadpoles raised in darkness do not function properly; exposure to light is required to allow them to become fully functional. We have no information about this problem in Sticklebacks, and this means that, in the absence of evidence with respect to either rods or cones in Sticklebacks, we must allow for the *possibility* that light — an environmental property - is required for the proper "programming" of part of the Stickleback's behaviour machinery. This brings me to another point which I consider important: the term "innate" whether applied to characters, or to differences, or to potentialities, or to developmental processes, is not the opposite of "learnt"; it is the opposite of "environment-induced".

These few considerations seem to me sufficient to conclude that the application of the adjective "innate" to behaviour *characters*, and to do this on the basis of eliminations of different kinds is heuristically harmful.

If I were to elaborate this further I should have to cross swords with my friend KONRAD LORENZ himself — both a pleasure and a serious task requiring the most thorough preparation — but this is not the occasion to indulge in swordplay, and I prefer to continue with my sketch of the procedure which seems to me more fruitful. This seems justified by the fact that the practice of ontogenetic research is not so much dependent on the background of the experimenter as the semantic and theoretical disagreements would lead one to suspect. The difference, for instance, between RIESS's (1950) and EIBL-EIBESFELDT's (1956) work on the ontogeny of nest building in rats is due just as much to a difference in the extent of knowledge of the Rat's normal behaviour as to the theoretical attitudes of the investigators.

A central issue in behaviour development studies seems to me the question raised by the fact that so many behaviour patterns can be said to be at the same time innate and learned, or partly innate and partly learned. EIBL-EIBESFELDT (1955) showed that nut-cracking in Squirrels consists of a series of component acts (manipulating, gnawing, and cracking) each of which develops in naive individuals which have not been able to practice; yet the adaptive integration of the acts into an efficient total pattern has to be learnt. Many similar cases are known, and it was a definite step forward when LORENZ (1937) coined the term „Instinkt-Dressur-Verschränkung" indicating that learning processes were often, so-to-speak, intercalated by non-learnt parts of a behaviour chain. This has given rise to the idea that, if one could only split up behaviour chains in smaller and smaller components, one could always reach a state where some components could be labelled as innate, others as learnt or acquired. I maintain that this may well be unhelpful, since many interactions with the environment which result in increased efficiency are *additive* to some machinery that was already functional. For instance, WELLS (1958) found that a naive young *Sepia* can perform the movements by which it catches *Mysis*, but that both the delay of the "attack" and the selectiveness of the response to stimuli decrease as a result of "having performed". In the sphere of conditioning — to mention a different form of environmental control of ontogeny — something similar is true: the animal is already selectively responsive (in other words it has an "IRM") before it has been conditioned; the conditioning changes a connection that was already present. The story of song development in Chaffinches as revealed by THORPE's work (1962) shows something very similar in the development of motor patterns: there is a definite pattern in the song of naive birds. I cannot see how, in view of such facts, it can be fruitful to look for innate and learned components, however small.

It seems to me that, if we return to a description of the phenomenon and the formulation of relatively simple questions, our course is laid out clearly.

The phenomenon (change in behaviour machinery) has to be described; the problem is, how are these changes controlled? As a first step one distinguishes between influences outside the animal and internal influences. External influences are usually detected by manipulating the environment during development, and the way in which such external agents influence the development can be studied along rather conventional lines, although it is obvious that, even in this relatively easier part of our task, we may well be in for some surprises, such as discovering that "rewards" may be of many more different types than known at present. One receives the first indications of *internal* control from demonstrations of the ineffectiveness of certain environmental properties, but the ultimate demonstration of internal control must come from direct interference with internal events. For instance, the development of successful ejaculation in Rats is, as BEACH and LEVINSON (1950) have shown, influenced by sex hormones which promote the growth of sensory papillae on the glans penis. Further insight into the internal control of growth of neural machinery is provided by the fascinating work of SPERRY (1959), particularly by his transplantation.of peripheral „Anlagen" before their innervation has been completed.

This general procedure, when applied at various levels of integration — to complex patterns, single acts, and even smaller components of the total behaviour machinery — seems to me much more fruitful than either basing a conclusion about innateness on elimination of part of the environmental properties, or proceeding on the assumption that all adaptedness of behaviour is acquired through interaction with the environment. It has been pointed out repeatedly (see PRINGLE 1951, LORENZ 1961) that there are two methods of "programming" the individual: the evolutionary trial-and-error-interaction with the environment which results in the specialisations of the genetic instructions, and the ontogenetic interaction between the individual and its environment — which, incidentally, takes the form of trial and error only where evolution has not given precise direction to the ontogenetic process.

I believe that such a procedure is in line with that widely applied in experimental embryology, which after all is the science concerned with exactly the same problem with reference to structure. And this takes me back to my starting point: by insisting on a biological approach LORENZ has influenced this aspect of Ethology as much as other aspects, even tough his evaluation of the part played by internal determinants may have been on the optimistic side.

I admitted above that in speaking of "the four problems of Biology" we apply a classification of problems which is pragmatic rather than logical. This is true of ontogeny in two respects at least. First, I have so far been speaking of the causation of ontogeny only, and it is clear that we must apply the question "what for?", the question of survival value to ontogeny as well. That is, we need to ask what the survival value is of the many different types of ontogenetic control that our analysis brings out. As yet we have only a hint of what is in store if we were to apply this question consistently. For instance, it is in some cases easy to see why the control in certain behaviour patterns is largely internal, and why in others interaction with the environment is advantageous. Thus a young Gannet, which has to jump off a high cliff, would be poorly off if he had to acquire the basic pattern of flight the way we acquire a skill such as writing. Similarly, the selective responsiveness to rival males in territorial species might well have to be unconditioned so that it can function at once when a male starts its first breeding cycle. Young song birds on the contrary, which begin by responding to a very wide range of objects when they start feeding independently and gradually learn to take only what has proved to be edible, are by this very "open-mindedness" able to adapt to many different habitats, and learn to select the most abundant food in each, however different this food may be in different habitats.

The study of ontogeny also overlaps, but in another way, with that of causation of cyclical or recurring behaviour in the adult. Some learning processes can occur all through the life of the individual, even though their impact decreases with age. (So do, of course, certain physiological changes, such as the formation of antibodies, or of pigmentation of the skin.) In this respect too ontogeny can be said to continue beyond the period of growth to maturity and the causation of the behaviour of the adult animal therefore grades into that of the phenomena usually classified under ontogeny; the distinction is partly one of the time scale involved. Yet there is sufficient justification to distinguish between the two sets of processes; as is obvious from the fact that one can say that a man is afraid of a flying plane "because he sees it" but also "because he has been bombed out as a child". The main point is to recognise that both statements may be true, that each covers part of the total causal chain involved, and that the question "what made him behave the way he did?" requires a complete answer in which both partial answers are contained.

### Evolution

The fact that behaviour is in many respects species-specific, and yet often similar in related species has been recognised by many workers before LORENZ, and the natural conclusion to be drawn from this, namely, that behaviour should be studied comparatively just as structures, with the ultimate aim of elucidating behaviour evolution, had also been drawn. WHITMAN's work (1898, 1919), and that of HEINROTH (1911), HUXLEY (1914, 1923) and VERWEY (1930), preceded LORENZ's contributions. While WHITMAN can be said to have concentrated on questions of homology or common descent, HUXLEY's interest was focussed on the task of testing the theory of natural selection. In a sense, however, all these important studies can be said to have been preliminary, preparatory to a concerted attack on problems of behaviour evolution which has gradually developed since LORENZ (1937) began to emphasise the need for systematic comparative studies.

In this field, too, research began in a rather intuitive way, guided by trends of thought which have been gradually made explicit and which have become increasingly similar to general evolutionary thinking. In this respect, too, Ethology is being incorporated into general Biology.

In some respects the evolutionary study of behaviour suffers from handicaps not met to that extent in that of structure. It needs no repeating that direct documentary evidence in the form of "fossil behaviour" is hardly available. The exploration of behaviour ontogeny as a tool does not seem to be very promising either. This, however, is not a serious handicap in view of the controversial nature of this tool in the study of structural evolution. In other respects ethologists are perhaps better off than students of structure: through their familiarity with the behaviour of many animals in their natural surroundings their attention has been drawn more readily to questions of survival value, and through these to a consideration of the effects of natural selection. When I say that ethologists were "better off" in this respect, I feel I should add in fairness to ourselves that this is due to our own efforts in creating a better opportunity; it is, in principle, easier to experiment on the survival value of structure than on that of behaviour, and the truth is that ethologists, by being in general good naturalists, deserved their good luck.

Evolutionary study has, of course, two major aims: the elucidation of the course evolution must be assumed to have taken, and the unravelling of its dynamics.

The first task is being pursued mainly through comparison of groups of closely related species. This limitation to closely allied forms is necessary because it is only here that conclusions about homology (i.e. common descent) can be drawn with any degree of probability. It is due to this restriction that what evidence we have applies to microevolution, particularly to adaptive radiation of relatively recent origin. As I have discussed elsewhere (TINBERGEN 1959), the trend here is to apply very much the same methods as those employed by taxonomists: we judge affinity by the criterion of preponderance of shared characters, particularly of those which we consider non-convergent. Once we have hardened the conclusion, often already reached by taxonomists, that a certain group must be monophyletic, we judge the degree of evolutionary divergence by the degree of dissimilarity of those characters that must be considered highly environment-resistant ontogenetically — we try to exclude from our material such differences as are the direct phenotypic consequence of different environments, such as an individually acquired darkening of external coloration under the immediate influence of a moist environment (which, of course, is very different from differences in *ability* to respond to the environment).

The comparative procedure has now been applied to a number of groups and one of the encouraging outcomes is that classifications based on behaviour taxonomy have, on the whole, corresponded very closely to the already existing classifications. Minor differences were found — but these often concerned matters which taxonomists considered not quite settled. I think it is worth emphasising this correlation between the two sets of results because it is again a striking justification of treating behaviour patterns as "organs".

The work on evolution *dynamics* can be said to consist of two major parts. First, the genetic control of species-specific behaviour, about which we know so much less than about that of species-specific structure, is now being studied with all the methods available in genetics; differences between species, subspecies and strains raised in identical environments are registered; the effect of mutations on behaviour are beginning to be explored, and controlled cross-breeding is being done. There seems little doubt as to the general outcome: individuals and populations differ as much in their hereditary behaviour "blueprints" as in their hereditary structural blueprints. The genetic variation on which natural selection can act is there.

The second major task is the study of the influence of selection on behaviour evolution. This task is being tackled in two different ways. One is the study of survival value of species-specific characters, the other is the direct application of a controlled selection pressure and its results over a series of generations.

The study of survival value receives its inspiration from the study of convergencies, and of divergencies within a taxonomic group; hypotheses about survival value are derived from these studies; they can be tested in experiment. The interpretation of the results is worth detailed scrutiny. When one finds that a certain characteristic has survival value — when it has been shown that various deviations from the norm lead to a lower success rate – one can draw one firm conclusion: one can say that one has demonstrated beyond doubt a selection pressure which prevents the species in its present state from deviating. One has really demonstrated the part played by selection in *stabilisation* of the present state. However, the conclusion that this same selection pressure must have been responsible in the past for the *moulding* of the character studied is speculative, however probable it often is. One can support such a conclusion by marshalling supplementary evidence, such as arguing that the environment which exerts the selection pressure must be assumed to have remained constant in this respect for a long time, or showing that the species is even now slightly variable according to area. Thus, E. CULLEN's demonstration of the adaptive nature of many of the Kittiwake's peculiarities (1957) can be used as a pointer to past selection pressures by arguing that the species has probably a fairly long history of cliff-nesting behind it. An example of the second line of argument is BLAIR's demonstration (1955) of the fact that the mating calls of two species of *Microhyla* are more distinct in the areas where they overlap (and where selection against crossbreeding must be assumed to have favoured inter-species distinctness) than in the areas where either species occurs alone.

All I have said above about the study of survival value for its own sake is relevant again here, but I should like to re-emphasize one point. For an assessment of what selection can be assumed to have contributed to the present state of species it is important to realise that selection rewards or penalises isolated bits of animals through rewarding or penalising animals, or breeding pairs, as wholes. Since studies of survival value show us that there are often direct contradictions between different selection pressures, the animal that survives best must be a compromise, and it must be one of our main tasks to try and find all the pressures — favourable and unfavourable alike — that can have affected any character we select for study. In general, we should not only try to pinpoint isolated selection pressures, but study their interaction as well.

In spite of the fact that we shall never be able to prove directly which contributions selection has made in the past, and that therefore any conclusion about the way interaction with the environment has moulded present-day species must remain tentative and, as such, different from conclusions drawn in the field of Physiology, Ontogeny and Survival Value, the ethologist feels that this is no reason to dismiss evolutionary study as just speculation. I believe that this developing branch of Ethology may well have effects on general biological thinking.

The direct application of selection pressures will, with increasing precision of description and measurements, give us an increasing amount of real demonstration of the potentialities of selection. With the growing trend towards experimentation it is important, however, to point out that even the most perfect experiment of this kind does not give us direct proof of what selection has done in the past. The interpretation of such experiments as contributions to evolution theory will always include an extrapolation: while they demonstrate what selection can do, the best they can tell us is that selection can have happened in the way demonstrated, and that the results obtained are not contradictory to what other indirect evidence has led us to suppose. They really deal merely with "possible future evolution", and only indirectly with past evolution. For instance, CROSSLEY's demonstration (1960) that 40 generations of antihybrid selection in partially interbreeding populations of ebony and vestigial mutants of *Drosophila melanogaster* change aspects of mating behaviour of both males and females of both populations does not directly prove that such selection *has* contributed to speciation, but it is in line with ideas developed before these experiments were done. This is, of course, not at all to belittle the relevance of such experiments, but merely to assign them their proper place in the total body of evidence.

## Conclusion

I have tried in this paper to give a sketch of what I believe modern Ethology to be about. I have perhaps given Ethology a wider scope than most practising ethologists would do, but if one reviews the various types of investigations carried out by people usually called ethologists, one is forced to conclude that the scope is in fact as wide as I have indicated. This sketch is not meant to be balanced or comprehensive; I have allowed myself to enlarge a little on special issues — on "bees in my bonnet" — such as the relations between Ethology and Physiology; the need to spend more effort on studies of survival value and methods to be employed in such studies; problems and methods of behaviour ontogeny; and the nature of arguments used in the study of evolution — all issues which require further discussion. I have also tried to assess KONRAD LORENZ's contribution to modern Ethology, and have argued that I consider his insistence that behaviour phenomena can, and indeed must, be studied in fundamentally the same way as other biological phenomena to be his major contribution. LORENZ can with justification be said to be the father of modern Ethology — even though he has had his forerunners; there is nothing amazing about every father having had a father.

The central point in LORENZ's life work thus seems to me his clear recognition that behaviour is part and parcel of the adaptive equipment of animals; that, as such, its short-term causation can be studied in fundamentally the same way as that of other life processes; that its survival value can be studied just as systematically as its causation; that the study of its ontogeny is similar to that of the ontogeny of structure; and that the study of its evolution likewise follows the same lines as that of the evolution of form. Moreover, in all these fields LORENZ has done concrete research which demonstrated the great heuristic value of his approach. Yet, although his concrete, factual contributions have been considerable, the impact he has made is due to his sketch of a type of approach, and to new and original hypotheses rather than to the experimental testing of such hypotheses. This is why recent changes of concepts and terminology, revisions of hypotheses, and the reporting of results which are sometimes different from LORENZ's earlier conclusions, have little relevance to the question of the value of his work.

One of the measures of this value which I will mention in passing is the fact that students of human behaviour are showing a growing interest in ethological methods.

Finally, I should like to touch briefly on a matter of terminology. It will be clear that I have used the word "Ethology" for a vast complex of sciences, part of which already have names, such as certain branches of Psychology and Physiology. This, of course, does not mean that I want to claim the name Ethology for this whole science, for this would be falsifying its history; the term really applies to the activities of a small group of biologists. What I have been at pains to develop is the thesis that we are witnessing the fusing of many sciences, all concerned with one or another aspect of behaviour, into one coherent science, for which the only correct name is *"Biology of behaviour"*. Of course this fusion is not the work of one man, nor of the small group called ethologists. It is the outcome of a widespread tendency to apply a more coherent biological approach, which has expressed itself in what may well have been quite independent developments within sciences such as Psychology and Neurophysiology. Among zoologists and naturalists, it is LORENZ who has contributed most to this development, and who has more than any other single person influenced these sister disciplines in this particular way. Finally, the comprehensive view of the aims of the biological study of behaviour has grown more rapidly in Ethology than in any of the other sciences. Yet, in view of the confused "public image" called up by the word Ethology it might well be advisable not to overdo the use of the word. What does seem to me to matter is the growing awareness of the fundamental unity of the Biology of Behaviour, and the realisation that "Ethology" is more than "Physiology of Behaviour", just as "Biology" is more than "Physiology".

## Zusammenfassung

Ich habe in diesem Aufsatz kurz anzudeuten versucht, was meiner Ansicht nach das Wesentliche in Fragestellung und Methode der Ethologie ist und weshalb wir in Konrad Lorenz den Begründer moderner Ethologie erblicken. Hierbei habe ich vielleicht das Arbeitsgebiet der Ethologie weiter gefaßt, als unter Ethologen gebräuchlich ist. Wenn man aber die vielartige Arbeit jener Forscher, die sich Ethologen nennen, ist man zu dieser weiten Fassung geradezu gezwungen. Ich habe in meiner Darstellung weder Vollständigkeit noch Gleichgewicht angestrebt und, um zur Fortführung des Gesprächs anzuregen, ruhig meine Steckenpferde geritten, vor allem das Verhältnis zwischen Ethologie und Physiologie, die Gefahr der Vernachlässigung der Frage der Arterhaltung, Fragen der Methodik der ontogenetischen Forschung, und Aufgaben und Methoden der Evolutionsforschung.

Bei der Einschätzung des Anteils, den Lorenz an der Entwicklung der Ethologie genommen hat und noch nimmt, habe ich als seinen Hauptbeitrag den bezeichnet, daß er uns gezeigt hat, wie man bewährtes „biologisches Denken" folgerichtig auf Verhalten anwenden kann. Daß er dabei an die Arbeit seiner Vorgänger angeknüpft hat, ist nicht mehr verwunderlich, als daß jeder Vater selbst einen Vater hat.

Insbesondere scheint mir das Wesentliche an Lorenz' Arbeit zu sein, daß er klar gesehen hat, daß Verhaltensweisen Teile von „Organen", von Systemen der Arterhaltung sind; daß ihre Verursachung genau so exakt untersucht werden kann wie die gleich welcher anderer Lebensvorgänge, daß ihr arterhaltender Wert ebenso systematisch und exakt aufweisbar ist wie ihre Verursachung, daß Verhaltensontogenie in grundsätzlich gleicher Weise erforscht werden kann wie die Ontogenie der Form und daß die Erforschung der Verhaltensevolution der Untersuchung der Strukturevolution parallel geht. Und obwohl Lorenz ein riesiges Tatsachenmaterial gesammelt hat, ist die Ethologie doch noch mehr durch seine Fragestellung und durch kühne Hypothesen gefördert. als durch eigene Nachprüfung dieser Hypothesen. Ohne den Wert solcher Nachprüfung zu unterschätzen — ohne die es natürlich keine Weiterentwicklung gäbe — möchte ich doch behaupten, daß die durch Nachprüfung notwendig gewordenen Modifikationen neben der Leistung des ursprünglichen Ansatzes vergleichsweise unbedeutend sind.

Nebenbei sei auch daran erinnert, daß eine der vielen heilsamen Nachwirkungen der Lorenzschen Arbeit das wachsende Interesse ist, das die Humanpsychologie der Ethologie entgegenbringt — ein erster Ansatz einer Entwicklung, deren Tragweite wir noch kaum übersehen können.

Am Schluß noch eine Bemerkung zur Terminologie. Ich habe hier das Wort „Ethologie" auf einen Riesenkomplex von Wissenschaften angewandt, von denen manche, wie Psychologie und Physiologie, schon längst anerkannte Namen tragen. Das heißt natürlich nicht, daß ich den Namen Ethologie für dieses ganze Gebiet vorschlagen will; das wäre geschichtlich einfach falsch, weil das Wort historisch nur die Arbeit einer kleinen Gruppe von Zoologen kennzeichnet. Der Name ist natürlich gleichgültig; worauf es mir vor allem ankommt, ist darzutun, daß wir das Zusammenwachsen vieler Einzeldisziplinen zu einer vielumfassenden Wissenschaft erleben, für die es nur einen richtigen Namen gibt: *„Verhaltensbiologie"*. Selbstverständlich ist diese synthetische Entwicklung nicht die Arbeit eines Mannes oder gar die der Ethologen. Sie ist die Folge einer allgemeinen Neigung, Brücken zwischen verwandten Wissenschaften zu schlagen, einer Neigung, die sich in vielen Disziplinen entwickelt hat. Unter den Zoologen ist es LORENZ, der hierzu am meisten beigetragen und zudem manche Nachbardisziplinen stärker beeinflußt hat als irgendein anderer. Ich bin sogar davon überzeugt, daß diese Einwirkungen auf Nachbarwissenschaften noch lange anhalten werden und daß die Verhaltensbiologie erst am Anfang ihrer Ontogenie steht.

## References

BEACH, F. A., and G. LEVINSON (1950): Effects of androgen on the glans penis and mating behavior of castrated male rats. J. Exp. Zool. 114, 159—168 • BEER, C. G. (1962): Incubation and nest-building behaviour of Blackheaded Gulls II: Incubation behaviour in the laying period. Behaviour 14, 283—305 • BLAIR, W. F. (1955): Mating-call and stage of speciation in the *Microhyla olivacea-M. carolinensis* complex. Evolution 9, 469—480 • BLEST, A. D. (1957): The function of eyespot patterns in the Lepidoptera. Behaviour 11, 209—256 • BLEST, A. D. (1960): The evolution, ontogeny and quantitative control of the settling movements of some New World Saturniid moths, with some comments on distance communication by Honey-bees. Behaviour 16, 188—253 • CROSSLEY, S. A. (1960): An experimental study of sexual isolation within the species *Drosophila melanogaster.* Anim. Behaviour 8, 232—233 • CROSSLEY, S. A. (Unpublished Doctor's thesis on the effect of anti-hybrid selection on sexual isolation within the species *Drosophila melanogaster.* Oxford) • CULLEN, E. (1957): Adaptations in the Kittiwake to cliff nesting. Ibis 99, 275—302 • CULLEN, E. (1961): The effect of isolation from the father on the behaviour of male Three-spined Sticklebacks to models. Tech. (Fin.) Rept. A F 61 (052)—29 U.S.A.F.R.D.C. • EIBL-EIBESFELDT, I. (1955): Über die ontogenetische Entwicklung der Technik des Nüsseöffnens vom Eichhörnchen *(Sciurus vulgaris L.)* Z. Säugetierk. 21, 132—134 • EIBL-EIBESFELDT, I. (1956): Angeborenes und Erworbenes im Nestbauverhalten der Wanderratte. Naturwiss. 42, 633—634 • GOETHE, F. (1937): Beobachtungen und Untersuchungen zur Biologie der Silbermöve auf der Vogelinsel Memmertsand. J. Ornithol. 85, 1—119 • GROHMANN, J. (1939): Modifikation oder Funktionsreifung? Z. Tierpsychol. 2, 132—144 • HAARTMAN, L. VON (1957): Adaptations in hole-nesting birds. Evolution 11, 339—348 • HASSENSTEIN, B. (1951): Ommatidienraster und afferente Bewegungs-Integration. Z. vgl. Physiol. 33, 301—326 • HASSENSTEIN, B. (1957): Über die Wahrnehmung der Bewegung von Figuren und unregelmäßigen Helligkeitsmustern. Z. vgl. Physiol. 40, 556—596 • HEINROTH, O. (1909): Beobachtungen bei der Zucht des Ziegenmelkers *(Caprimulgus europaeus* L.) J. Ornithol. 57, 56—83 • HEINROTH, O. (1911): Beiträge zur Biologie, namentlich Ethologie und Psychologie der Anatiden. Verh. V. Internation. Ornith. Kongr. Berlin • HEINROTH, O. und M. (1928): Die Vögel Mitteleuropas. Berlin • HINDE, R. A. (1958): The nest-building behaviour of domesticated canaries. Proc. Zool. Soc. Lond. 131, 1—48 • HINDE, R. A. (1959): Unitary drives. Anim. Behaviour 7, 130—141 • HINDE, R. A. (1962): Temporal relations of broodpatch development in domesticated canaries. Ibis 104, 90—97 • HINDE, R. A., and R. P. WARREN (1961): Roles of the male and the nest-cup in controlling the reproduction of female canaries. Anim. Behaviour 9, 64—67 • HOLST, E. VON (1937): Vom Wesen der Ordnung im Zentralnervensystem. Naturwiss. 25, 625—631, 641—647 • HOLST, E. VON, und H. MITTELSTAEDT (1950): Das Reafferenzprinzip. Naturwiss. 37, 464—476 • HOLST, E. VON, und U. VON ST. PAUL (1960): Vom Wirkungsgefüge der Triebe. Naturwiss. 47, 409—422 • HOOGLAND, R., D. MORRIS and N. TINBERGEN (1957): The spines of sticklebacks *(Gasterosteus* and *Pygosteus)* as means of defence against predators *(Perca* and *Esox).* Behaviour 10, 205—236 • HUXLEY, J. S. (1914): The courtship habits of the Great Crested Grebe *(Podiceps cristatus);* with an addition to the theory of sexual selection. Proc. Zool. Soc. Lond. 1914, 491—562 • HUXLEY, J. S. (1923): Courtship activities in the Red-throated Diver *(Colymbus stellatus* Pontopp); together with a discussion on the evolution of courtship in birds. Jour. Linn. Soc. 35, 253—291 • KETTLEWELL, H. B. D. (1955): Selection experiments on industrial melanism in the Lepidoptera. Heredity 9, 323—342 • KETTLEWELL, H. B. D. (1956): Further selection experiments on industrial melanism in the Lepidoptera. Heredity 10, 287—301 • KLOMP, H. (1954): De Terreinkeus van de Kievit, *Vanellus vanellus* (L.). Ardea 42, 1—140 • KNOLL, M. (1953): Über das Tages- und Dämmerungssehen des Grasfrosches nach Aufzucht in veränderten

Lichtbedingungen. Z. vgl. Physiol. **35**, 42—67 • LEHRMAN, D. S. (1953): A critique of KONRAD LORENZ's theory of instinctive behavior. Quart. Rev. Biol. **28**, 337—363 • LEHRMAN, D. S. (1955): The physiological basis of parental feeding behaviour in the Ring Dove *(Streptopelia risoria).* Behaviour **7**, 241—286 • LEHRMAN, D. S. (1961): Hormonal regulation of parental behaviour in birds and infrahuman mammals. in: W. C. YOUNG (Ed.): Sex and Internal Secretion, Baltimore • LINDAUER, M. (1961): Communication among Social Bees. Cambridge Mass. • LORENZ, K. (1935): Der Kumpan in der Umwelt des Vogels. J. Ornithol. **83**, 137—213, 289—413 • LORENZ, K. (1937): Über die Bildung des Instinktbegriffes. Naturwiss. **25**, 289—300, 307—318, 324—331 • LORENZ, K. (1955): (Discussion remark on p. 77 in: Group Processes, First Josiah Macy Jr Foundation Conference. New York.) • LORENZ, K. (1950): The comparative method in studying innate behaviour patterns. S.E.B. Symposia **4**, 221—269 • LORENZ, K. (1961): Phylogenetische Anpassung und adaptive Modifikation des Verhaltens. Z. Tierpsychol. **18**, 139—187 • MOSEBACH-PUKOWSKI, E. (1937): Über die Raupengesellschaften von *Vanessa io* und *Vanessa urticae.* Z. Morph. Oekol. Tiere **33**, 358—380 • NIELSEN, E. T. (1958): The Method of Ethology. Proc. 10th Internat. Congress of Entomol. **2**, 563—565 • PRINGLE, J. W. S. (1951): On the parallel between learning and evolution. Behaviour **3**, 174—216 • RIESS, B. F. (1950): The isolation of factors of learning and native behavior in field and laboratory studies. Ann. N.Y. Acad. Sci. **51**, 1093—1102 • ROWELL, C. H. F. (1961): Displacement grooming in the Chaffinch. Animal Behaviour **9**, 38—64 • RUITER, L. DE (1952): Some experiments on the camouflage of stick caterpillars. Behaviour **4**, 222—232 • SAUER, F. (1954): Die Entwicklung der Lautäußerungen vom Ei ab schalldicht gehaltener Dorngrasmücken *(Sylvia c. communis* Latham) im Vergleich mit später isolierten und wildlebenden Artgenossen. Z. Tierpsychol. **11**, 10—93 • SEVENSTER, P. (1961): A causal analysis of a displacement activity. Behaviour Suppl. **9** • SPERRY, R. W. (1959): The growth of nerve circuits. Sci. American **201**, 68—75 • THORPE, W. H. (1962): Bird Song. Cambridge • TINBERGEN, N. (1940): Die Übersprungbewegung. Z. Tierpsychol. **4**, 1—40 • TINBERGEN, N. (1951): The Study of Instinct. Oxford • TINBERGEN, N. (1959): Comparative studies of the behaviour of gulls; a progress report. Behaviour **15**, 1—70 • TINBERGEN, N., and A. C. PERDECK (1950): On the stimulus situation releasing the begging response in the newly hatched Herring Gull chick *(Larus a. argentatus* Pontopp.). Behaviour **3**, 1—38 • TINBERGEN, N., G. J. BROEKHUYSEN, F. FEEKES, J. C. W. HOUGHTON, H. KRUUK and E. SZULC (1962): Egg shell removal by the Black-headed Gull, *Larus ridibundus* L.; a behaviour component of camouflage. Behaviour **19**, 74—118 • VERWEY, J. (1930): Die Paarungsbiologie des Fischreihers. Zool. Jahrb. Allg. Zool. Physiol. **48**, 1—120 • WELLS, M. J. (1958): Factors affecting reactions to *Mysis* by newly hatched *Sepia.* Behaviour **8**, 96—111 • WHITMAN, Ch. O. (1898): Animal Behavior. Woods Hole • WHITMAN, Ch. O. (1919): The Behavior of Pigeons. Carnegie Inst. Wash. Publ. **257**, 3, 1—161 •

TINBERGEN, N. (1963): On aims and methods of ecology. Z. Tierpsychol., **20**, 410—433. Reprinted with kind permission from Blackwell Publishing Ltd.

JERRY A. HOGAN AND JOHAN J. BOLHUIS

## 2

# Tinbergen's four questions and contemporary behavioral biology

### FOUR QUESTIONS ON BEHAVIOR

Tinbergen's goal in his paper "On aims and methods of Ethology" (1963, this volume) was to define the field of ethology and demonstrate the close affinity between ethology and the rest of biology. He also wanted to show the ways in which ethology was similar to, and different from, other life sciences. His definition was simple: Ethology is "*the biological study of behaviour.*" He elaborated his meaning of "biological study" by stating that "the biological method is characterized by the general scientific method, and in addition by the kind of questions we ask (…)." The questions were those of causation, survival value, and evolution, as stated by Huxley (1914), plus that of ontogeny, added by Tinbergen. In the body of the paper, Tinbergen discusses and illustrates each question using examples from then-current ethological work. Further, he notes that most ethologists at that time had concentrated their efforts on asking causal questions, and pleads for more studies that ask the other questions as well. It was his opinion that answers to all four questions would lead to a much greater understanding of behavior. One other point is important to mention: Tinbergen's statement that his classification of problems is pragmatic rather than logical (Tinbergen, 1963, p. 426) may have contributed to some of the confusions we discuss below, as well as to extensions of these questions discussed in later chapters of this book.

First, it is important to realize what the four questions – or the four whys as they are also known – are about. Many authors consider an understanding of the four questions a crucial prerequisite for the study of animal behavior (e.g., Manning and Dawkins, 1995; Barnard, 2004; Bolhuis and Giraldeau, 2005). For instance, in the introductory chapter of their textbook, Bolhuis and Giraldeau (2005) state that "Tinbergen's

*Tinbergen's Legacy: Function and Mechanism in Behavioral Biology*, Johan J. Bolhuis and Simon Verhulst (eds.). Published by Cambridge University Press. © Cambridge University Press 2008.

analysis is so important that we would say that you cannot really understand animal behavior if you do not also understand the meaning of Tinbergen's four questions." The authors illustrate the four whys with the example of a male songbird singing at dawn. The question "why does this bird sing?" is not very useful, as it can have at least four different meanings. A student of birdsong has to make clear which of Tinbergen's four whys is being addressed. The first of the four questions concerns causation: what causes the bird to sing? Another way of asking this is: what are the mechanisms underlying the male's singing behavior? These mechanisms involve the "machinery" that operates within the animal and which is responsible for the production of behavior. The topics include the stimuli or triggers of behavior whether they be internal or external, the way in which behavioral output is guided, factors that stop behavior and the like. The problem of the *causation* of behavior is sometimes called motivation, a topic discussed at length by Hogan in Chapter 3 of the present volume. The second question is about development. How did the singing behavior of the bird come about in the lifetime of an animal? Research has shown that a male songbird does not sing immediately after it has hatched from the egg, but that it develops a song in a process that involves learning (see Bolhuis, present volume and Bolhuis and Gahr (2006) for a recent review). Such questions that concern *development* of behavior – Tinbergen termed this "ontogeny" – will be discussed explicitly by Crews and Groothuis (Chapter 4) and Hogan and Bolhuis (Chapter 5) in the present volume. The third question has to do with function, or what Tinbergen termed "survival value." What is the function of the bird singing? This question has to do with the consequences of singing for the singer's fitness. Does singing help the bird keep intruding males away from his nest? Or, alternatively, does it serve to attract females? The question of function is discussed in detail by Cuthill in Chapter 6 of the present volume. The fourth question concerns evolution: how did this behavior come about in the course of evolution? Was the current song derived from a more primitive song sung by the species' ancestors? Behavior does not leave many fossils behind and so the study of its evolutionary history requires the development of special methods, as discussed by Ryan in Chapter 7 of the present volume.

## TINBERGEN AND THE RISE OF ETHOLOGY

It is useful to consider the intellectual climate in which Tinbergen first formulated the four questions. By the middle of the 20th century, there were at least three major groups of scientists studying the behavior of animals: the comparative psychologists, mostly in North America; the ethologists, originally mostly in Germany and the Netherlands, but soon thereafter also in Great Britain and other parts of Europe; and the evolutionary biologists, originally mostly in Britain, but very soon thereafter also in North America and Europe. There was much overlap between these groups, but the underlying philosophy of each sprang from different origins: the comparative psychologists were rooted in the laboratory traditions of the German psychologists of the late 19th century, but adapted to the pioneering traditions of North America. The early ethologists were largely field biologists interested in natural history, and the evolutionary biologists were concerned to show how the observed varieties of behavior could be understood according to Darwinian principles of evolution through natural selection. Both the comparative psychologists and the ethologists were primarily interested in the behavior of individual animals, whereas the evolutionary biologists were primarily interested in the selection pressures that led to behavioral adaptations and ultimately to the origin of species. In spite of these differences, all three groups had in common that, as scientists, they were regarded as second-class citizens in comparison to the physicists, chemists, and physiologists. It is in this context that Tinbergen (1963) formulated his four questions in an attempt to unify the field of ethology. (See Röell, 2000, and Burkhardt, 2005, for more background information on the early approaches to the study of animal behavior.)

In this chapter, we will first discuss Tinbergen's formulation of the questions. We then follow the vicissitudes this formulation underwent in the hands of different scientists. Finally, we indicate which distinctions we feel will be most important for researchers in the future.

OTHER FORMULATIONS AND INTERPRETATIONS

### The concept of cause

Tinbergen used the term "causation"– his first question – to mean the immediate effects that external and internal factors have on the occurrence of behavior. (For an extended discussion of Tinbergen's conception of causation, see Hogan, this volume.) This is also similar to the dictionary definition of the word "cause": "that which produces an effect, result, or consequence" (*American Heritage Dictionary* – AHD,

1969). The dictionary definition, however, is much broader than Tinbergen's usage, and confusion can arise when the meaning of the word is simply assumed. For example, many authors have gone back to Aristotle in order to arrive at a logical classification of problems in behavior (e.g., Hopkins and Butterworth, 1990; Dewsbury, 1992, 1999; Hogan, 1994; Killeen, 2001; Short, 2002). Aristotle, in translation, spoke of the four "causes" of any natural phenomenon: material, efficient, formal, and final causes. With respect to behavior, these causes would be the nervous system and muscles necessary for expression of the behavior (material), the external and internal factors that activate the behavior (efficient), the organization of these factors that determines the form of the behavior (formal), and the outcome or purpose of the behavior (final). Of these, only the efficient cause corresponds to Tinbergen's usage. A more complete discussion of the problems these various uses of the word raise can be found in Hogan (1994).

## The concept of function

Tinbergen's second question – survival value – asks what is the function of a behavior for the survival of the animal. (For an extended discussion of Tinbergen's conception of survival value or function, see Hinde, 1975, and Cuthill, this volume.) Once again, the AHD shows that the word "function" has a wider meaning: "the natural or proper action for which a person, office, mechanisms, or organ is fitted or employed." (See also Wouters, 2003, 2005). It is the words "natural" and "fitted" that lead to the most controversial issues. These words imply to some that the behavior under consideration was somehow designed to fit its current function. Aristotle's "final cause" is an example of this interpretation. It is also an example of teleology, which implies that the goal of behavior (what it is designed for) has become its efficient cause. *The outcome of behavior can never determine its occurrence.* At best, the outcome of behavior can be one of the determining (causal) factors of future occurrences of similar behavior, either through motivational changes or learning during the lifetime of an individual, or through natural selection during evolution. Tinbergen (1963) himself discusses the problems that teleology gives rise to in the study of behavior, as have many other authors (e.g., Pittendrigh, 1958; Mayr, 1961, 1988; Hopkins and Butterworth, 1990; Hogan, 1994; Short, 2002; Bolhuis, this volume).

## The concept of proximate and ultimate causes

Evolutionary biologists have approached the classification of problems in biology somewhat differently from Tinbergen. Baker (1938) introduced the terms "proximate causation" and "ultimate causation" and Mayr (1961, p. 1503) used these terms to distinguish between the problems of interest for the functional biologist (proximate causes: "the immediate set of causes") and for the evolutionary biologist (ultimate causes: "causes that have a history and that have been incorporated into the system through many thousands of generations of natural selection"). The proximate causes of Mayr correspond almost exactly with Tinbergen's use of the word "causation"; but the ultimate causes of Mayr correspond only partially to Tinbergen's problems of survival value and evolution, even though most current authors treat them as the same (for further discussion, see Cuthill, this volume).

It is Mayr's terminology that has become the standard in the fields of behavioral ecology and evolutionary psychology, but, as with Tinbergen's terms, the words "proximate" and "ultimate" have many meanings, and various interpretations of these terms soon appeared. Francis (1990) published a critique of the distinction, especially of the term "ultimate." Francis argued that explanations referring to ultimate causes typically emerge from functional analyses, but that functional analyses do not identify causes of any kind resembling those of proximate explanations. He also pointed out that the term "ultimate cause" implies that functional analyses are somehow superordinate to those involving proximate causes. He concluded that ultimate causes are neither ultimate nor causes. Similar arguments have been made more recently by Bolhuis and Macphail (2002) and by Bolhuis (this volume). Bolhuis argues against the use of the term "ultimate mechanisms" (Pravosudov and Clayton, 2001; Healy *et al.*, 2005) because it implies that evolution or natural selection is a causal factor in behavior or cognition. He argues that natural selection can be seen as the causal factor in the historical *process* of evolution, but not a cause of the behavioral, cognitive, or neural phenotype.

Mayr (1993) responded to Francis's critique quite vehemently. We cannot discuss all the relevant issues here, but one general point is worth making. Mayr chided Francis for going to the dictionary for the definition of the word ultimate: "It has always seemed to me a dubious procedure to search in a dictionary for authoritative information on a scientific term" (p. 63). We agree that the dictionary is not the place to look for an authoritative scientific definition of a term, but the problem is that most people who read a word will understand that word in its usual meaning, whether an author intends it that way or not. Mayr (1993, p. 94) provides a scientific definition of "ultimate causations": the laws "which cause changes in the DNA of genotypes." We do not disagree with this formulation because those laws must be the ones that cause mutations and genetic recombination which cause the variation upon which natural selection acts. Unfortunately, this meaning is not what most authors of papers in behavioral ecology and evolutionary psychology have in mind when they use the term: "Ultimate causation concerns adaptive significance" (Daly and Wilson, 1978, p. 9), a meaning Mayr specifically eschewed. Note also that Mayr's 1993 definition is quite different from his 1961 definition cited above. Dewsbury (1999) provides an excellent analysis of how these terms have been used and misused by behavioral scientists.

### The relation of cause and function

As Hogan (1994) pointed out, cause and function are logically independent concepts, a view that is endorsed by Bolhuis (2005, present volume). This is so because the cause of evolution is the mutation or recombination of genes that bring about variability among individuals (cf. Mayr's (1993) definition above). Natural selection is the consequence or outcome of the fact that some individuals are more successful than others. That is, in some particular environment, individuals with a certain genotype will be more successful than individuals with a different genotype. However, that environment does not cause the successful genotype to appear; it only maintains that genotype once it has been caused by other (causal) factors.

Nonetheless, as Bolhuis (Chapter 9, this volume) discusses, functional considerations can sometimes provide clues about possible causal mechanisms. An example is provided by recent work on behavioral syndromes (personality) in various species of animals (Dingemanse and Réale, 2005; Dingemanse et al., 2007). This work shows that behavioral correlations were probably caused by pleiotropic effects of genes affecting multiple behaviors, and that this causal mechanism only exists in environments where selection favors these correlations. Hence, although cause and function are logically independent concepts, one can sometimes predict function from cause and vice versa because of the process of evolution.

### The problem of levels of analysis

The phrase "levels of analysis" has been used in two completely different ways by behavioral scientists. On the one hand, many evolutionary biologists and psychologists use the phrase as a synonym for Tinbergen's four questions. Daly and Wilson (1978, p. 11), for example, in a section entitled "levels of behavioral explanation," mention Tinbergen's "four great problems ... cause, function, development, and evolution. They correspond to our proximate causation, ultimate causation, ontogeny, and phylogeny." Similarly, Sherman (1988, p. 616) says: "In summary, there are four different levels of analysis: evolutionary origins, functional consequences, ontogenetic processes and mechanisms." This terminology continues to be used to the present (e.g., Barnard, 2004; Ryan, this volume). Although all behavioral scientists would agree that there are different *kinds* of explanations, the use of the word "level" is unfortunate. The AHD defines "level" as "relative position or rank on a scale." It is not possible to arrange or rank Tinbergen's questions on any reasonable scale, although the use of the word "ultimate" may suggest to some that ultimate explanations are somehow higher or more important than other kinds of explanations. This aspect of the word "ultimate" is also discussed by Francis (1990) and Dewsbury (1999).

The phrase "levels of analysis" has been used in a second way by Tinbergen, Hinde (1970), and Hogan (1994). In this case, "level" is the correct word, and the scale goes from genes to biochemistry to physiology to behavior to social behavior. In his concluding remarks to the section on causation, for example, Tinbergen (1963, p. 416) states: "As far as the study of causation of behaviour is concerned the boundaries between these fields are disappearing, and we are moving fast towards one Physiology of Behaviour, ranging from behaviour of the individual and even of supra-individual societies all the way down to Molecular Biology." Although Tinbergen is here recommending that we abolish the distinctions between levels, another of his examples shows what confusions can result if this is done. He points out (p. 414) that "... the more sophisticated ethologist is fully aware of the fact that the term "Innate Releasing Mechanism" refers to a type of function, of achievement, found in many different animals; and that different animal types may well have convergently achieved [this] by entirely different mechanisms ...". The confusion is that, at the neurophysiological level, "releasing mechanism" is a functional concept as Tinbergen states, while at the behavioral level it is a causal concept: It is postulated that activation of a particular releasing mechanism is one of the causal factors for the occurrence of a certain behavior. Many years ago, von Holst and von St. Paul (1960) noted that it is necessary to use level-adequate concepts when studying complex behavior, and this point is discussed more extensively by Hogan (1994, this volume).

### The question of ontogeny

Ontogeny is defined by the AHD as "the course of development of an individual organism," which is clearly an issue that needs to be investigated by a science of behavior. Tinbergen added ontogeny to his list of questions to a large extent because ethology had been strongly criticized for ignoring questions of development (Lehrman, 1953). As Crews and Groothuis (this volume) point out, however, Tinbergen formulated the study of behavioral development as the study of causal mechanisms. They continue: "In fact, all of [his] other ... questions ... can and should be applied to the study of behavioural development." Viewed in this light, ontogeny is clearly a different kind of question than cause and function. We return to this issue in our conclusions.

### The question of evolution

The general definition of evolution in the AHD is: "A gradual process in which something changes into a significantly different, especially more complex or more sophisticated, form." This definition can just as easily apply to ontogeny as to the origin of species. A more technical word that is generally used for the latter changes is *phylogeny*: "The evolutionary development of any species of plant or animal." Thus, the difference between ontogeny and phylogeny is that the former deals with changes in an individual organism during its lifetime, while the latter deals with changes in species (or groups of organisms) over the course of generations. It follows that ontogeny and phylogeny refer to different levels of analysis, in the good sense as discussed above. At both levels, causal and functional questions can and should be asked. But, as Ryan (this volume) discusses, "phylogenetic methods" have been applied to both individual behavior and species behavior. And, as we have seen above, an outcome or function at one level (e.g., the neurophysiological level) can be a cause at the next higher level (e.g., the behavioral level). This means that special care must be taken with our vocabulary when discussing these issues.

## SUMMARY AND CONCLUSIONS

Tinbergen formulated his four questions (causation, survival value, ontogeny, and evolution) in the context of his view of ethology in the early 1960s. Since then, these questions have served as a useful framework for many workers in the field of animal behavior. However, as we have seen, the interpretation of these questions has varied among authors, and this has often led to confusion. Further, many workers have conflated Tinbergen's questions with the distinction between proximate and ultimate factors made by many behavioral biologists (cf. Cuthill, this volume), and this has led to still further confusion, as well as to a certain amount of acrimony (Bolhuis and Macphail, 2001; Bolhuis, this volume). We feel that Lehrman (1970, pp. 18–19) addressed these issues correctly, when he stated that "When opposing groups of intelligent, highly educated, competent scientists continue over many years to disagree, and even to wrangle bitterly, about an issue which they regard as important, it must sooner or later become obvious that the disagreement is not a factual one, […]. If this is, as I believe, the case, we ought to consider the roles played in this disagreement by semantic difficulties arising from concealed differences in the way different people use the same words, or in the way the same people use the same words at different times; by differences in the concepts used by different workers (i.e., in the ways in which they divide up facts into categories); and by differences in their conception of what is an important problem and what is a trivial one, or rather what is an interesting problem and what is an uninteresting one."

We believe that many of the issues mentioned above illustrate Lehrman's point. Various attempts to provide a logical classification of problems (e.g., Dewsbury, 1992; Hogan, 1994) have had little or no influence on the field, and it is unlikely that other attempts will be more successful. Further, it is certain that authors will continue to use words such as "cause," "function," "ultimate," "levels," and "evolution," and that different authors will use these words in ways that have different meanings. The best that can be hoped for is that authors and readers both realize that this is the case, and try to be as clear as possible. It is also certain that scientists will continue to work on problems that they find important and interesting, but that other scientists may find trivial, uninteresting, or old-fashioned. It seems to be human nature to feel that if one "knows the truth," everyone else should agree. Human nature is unlikely to change, but it is possible to be more open-minded and accepting of different interests and points of view. However, we also believe that there are some *factual* matters that one can conclude from our discussion. Of these, the distinction between cause and consequence is paramount. Further, by whatever name you wish to call them, there are different *kinds* of questions one can ask about behavior, and addressing them all will lead to a better understanding. And finally, the concepts used to analyze behavior should be appropriate to the aspect of behavior (e.g., individual or group) being investigated.

JERRY A. HOGAN

3

# Causation: the study of behavioral mechanisms

This chapter provides an overview of studies on the causal analysis of behavior systems in the 40 years since Tinbergen (1963) published his views on the aims and methods of ethology. I begin with some comments on Tinbergen's conception of causation. It is then noted that, while causal work investigating the neural, hormonal, and genetic bases of behavior is flourishing, work being conducted at a strictly behavioral level of analysis has declined greatly in the past 40 years. Nonetheless, most recent research on animal cognition and applied ethology is still being carried out at a behavioral level of analysis and examples of both types of research are presented: memory mechanisms of food-storing birds and decisions of spider-eating jumping spiders as well as feather pecking in fowl and animal welfare issues are all briefly discussed. Finally, I discuss the similarities between neural network modeling and early ethological models of motivation, and then show how a modern version of Lorenz's model of motivation can account for current research findings on dustbathing in chickens and sleep in humans. I conclude that valuable information can still be obtained by research at a behavioral level of analysis.

## CONCEPTION OF CAUSATION

Tinbergen uses the word "causation" to refer exclusively to what many behavioral biologists currently call "proximate causation": the immediate effects that external and internal factors have on the occurrence of behavior. He begins his analysis of causation by referring to three statements made by Lorenz (1937), which he then proceeds to discuss.

The first statement is that animals can be said to *possess* behavior characteristics just as they *possess* certain structural and physiological characteristics. Tinbergen argues that this formulation allowed ethology

*Tinbergen's Legacy: Function and Mechanism in Behavioral Biology*, Johan J. Bolhuis and Simon Verhulst (eds.). Published by Cambridge University Press. © Cambridge University Press 2008.

to rid itself of subjectivism, anthropomorphism, and teleology (giving function as a proximate cause). It is not clear why this particular formulation had these effects, especially since other formulations such as those of the behaviorists (Skinner, 1953) had similar effects. It is likely that most students of behavior at that time were intent on proving that they were scientists. However, an important fact pointed out by Tinbergen is that "throughout biology we tend to classify, and hence to give names, on the basis of criteria of common function." This can be a major stumbling block to causal analysis. One example is the *releasing mechanism (RM)*, a concept that refers to a type of function found in many different animals, and which may have convergently achieved this function by entirely different genetic and neural mechanisms. It is important here to realize that, at the behavioral level, the releasing mechanism is a causal concept in that activation of the RM causes a particular behavior to occur, while at the physiological level it is a functional concept, as Tinbergen pointed out. A second type of difficulty is "our habit to coin terms for major functional units such as nest building, fighting, or sexual behavior and treat them as units of mechanism." For example, sexual behavior, defined functionally, would be all behavior that serves a reproductive function. Defined causally, sexual behavior would be all behavior controlled by a specific sexual mechanism. The causal definition would exclude many displays seen during courtship; these displays may be necessary for successful reproduction, but they are believed, in many cases, to be controlled by aggressive and escape mechanisms (Tinbergen, 1952; Baerends, 1975; Groothuis, 1994).

The fact that behavior mechanisms can be defined both causally and functionally and can be functional at one level of analysis, but causal at a higher level means that one should be very careful to define one's units of analysis. I have suggested that this becomes easier if one makes a clear distinction between structure and causation (Hogan, 1994, 2001). I have proposed perceptual, central, and motor mechanisms as the basic structural units of behavior. These entities are viewed as corresponding to structures within the central nervous system. They consist of an arrangement of neurons (not necessarily localized) that acts independently of other such mechanisms. Perceptual mechanisms analyze incoming sensory information and solve the problem of stimulus recognition. An example is the releasing mechanism discussed above. The motor mechanisms are responsible for coordinating the neural output to the muscles, which results in recognizable patterns of movement. The central mechanisms coordinate the perceptual and motor mechanisms, and provide the basis for an animal's mood or internal state. These units are

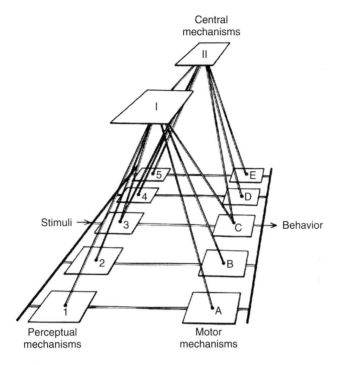

Fig. 3.1. Conception of behavior systems. Stimuli from the external world are analyzed by perceptual mechanisms. Output from the perceptual mechanisms can be integrated by central mechanisms and/or channeled directly to motor mechanisms. The output of the motor mechanisms results in behavior. In this diagram, central mechanism I, perceptual mechanisms 1, 2, and 3, and motor mechanisms A, B, and C form one behavior system; central mechanism II, perceptual mechanisms 3, 4, and 5, and motor mechanisms C, D, and E form a second behavior system. Systems 1-A, 2-B, and so on can be considered less complex behavior systems. (From Hogan, 1988.)

called behavior *mechanisms* because their activation results in an event of behavioral interest: a particular perception, a specific motor pattern, or an identifiable internal state. Behavior mechanisms can be connected with one another to form larger units called behavior systems, which correspond to the level of complexity indicated by terms such as feeding, sexual, and aggressive behavior. The organization of the connections among behavior mechanisms determines the nature of the behavior system. Thus, a behavior system can be considered a description of the structure of behavior (see Fig. 3.1). As I show below, Tinbergen does not always make such clear distinctions and this has led to some confusion about the meaning of causal analysis.

The second statement of Lorenz is that what ethologists call behavior is vastly more complex than the types of movements usually studied by physiologists; and the third statement is that behavior is controlled by internal causal factors to a greater extent than most physiologists believed at that time. These discrepancies led workers in each field to develop and use their own concepts. Once again, we are dealing with a problem of level of analysis. In discussing this issue, Tinbergen says that he believes "that it is doing our science a great deal of harm to impose boundaries between it and physiology where there are none ..." It is difficult to disagree with this statement, but, unfortunately, this point of view also implies that the proper study of the causation of complex behavior is always reductive. In fact, Tinbergen states specifically that the only acceptable name for the study of causation of animal movement with respect to all levels of integration would be "*Physiology of Behaviour.*" I believe that it is both desirable and necessary to use different concepts for each level of analysis: physiological concepts are much more appropriate for understanding the causation of muscle contractions and other reflex-type behaviors, than they are for understanding the causation of the behavior systems studied by many ethologists. As von Holst and von St. Paul (1960) point out, it is necessary to use level-adequate concepts ("niveau-adäquate Terminologie") when studying complex behavior: "The fact that particular behaviour elements can belong as components to completely different drives (and moods) is theoretically important. It forces us in every case to search for and to *name the highest integrated unit* activated by the stimulus field. For if the terminology does not reflect the synthesizing capacity of the CNS, the reconstruction of the functional organization is made impossible from the beginning." (1960, p. 17) I return to this point at the end of this chapter.

### STUDIES OF CAUSATION

If we examine what has happened to studies of causation since Tinbergen's paper, two major trends can be seen. First, many studies have become more molecular with respect to methods and questions asked. Neuroethology and behavioral endocrinology have become fields in their own right, with their own scientific meetings, specialized journals, and textbooks (e.g., Nelson, 1999; Carew, 2000). Much progress has been made in both fields. In neuroethology, there are elegant studies of the neural correlates of electrolocation and jamming avoidance in electric fish (Heiligenberg, 1977), of prey-catching and avoidance behavior of amphibians (Ewert, 1997), of bird song development (Nottebohm, 2002), and of the coding of auditory space in barn owls (Konishi, 2003), to name only a few. In behavioral endocrinology, the pioneering studies of Beach (1948) on the role of hormones in male rat sexual behavior and of Lehrman (1965) on the role of hormones in the reproductive behavior of ring doves have been followed up by studies of the effects of neuroendocrinological factors on parental behavior in rats and humans (Fleming and Blass, 1994), on reproductive behavior of lizards and snakes (Crews and Silver, 1985; Crews and Groothuis, this volume), and on bird song behavior and development (Ball *et al.*, 2002), among many others. Behavioral genetics is a third area that is currently making rapid progress. Using both traditional and more recent molecular methods developed by geneticists, Marla Sokolowski (2001) and her colleagues have been studying how specific genes affect many of the complex behavior patterns of the fruit fly *Drosophila melanogaster*. Derk-Jan Dijk and his colleagues have been studying the effects of specific genes on human sleep. They have found that polymorphism in the circadian clock gene PER3 predicts sleep structure and waking performance in human subjects (Viola *et al.*, 2007) It is noteworthy that studies in all these fields are seldom published or referred to in the mainstream ethological journals, *Animal Behaviour, Behaviour*, and *Ethology*.

The second major trend in studies of causation at the level of the individual animal, or the behavior system level, is that such studies have become relatively much less numerous. In 1963, Tinbergen argued that more effort should be expended on studies of survival value. In some ways recent history shows this effect has been achieved. I have looked through the table of contents of research articles published in *Animal Behaviour* at 10-year intervals since 1963, and, primarily on the basis of title, have categorized each article as being primarily a study of causation or survival value (function). The few studies on ontogeny were grouped with causation and the few studies of evolution were grouped with function. The results are shown in Fig. 3.2. The graph shows clearly that studies of causation remained the primary interest of ethologists until at least 1973, but have almost disappeared since then. The continued high number of studies of causation through the early 1970s suggests that Tinbergen's exhortation to divert more attention to studies of function was not actually heeded. What is almost certainly the case is that the publication of Wilson's (1975)

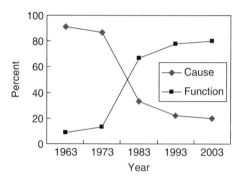

Fig. 3.2. The percentage of research articles published at 10-year intervals in *Animal Behaviour* that are primarily studies asking causal questions or functional (survival value) questions.

book *Sociobiology* was the major impetus behind the switch of attention to functional questions.

In spite of the massive shift of emphasis in the ethological journals, studies of causal questions at the level of the individual have continued to be done, but they have been published mostly in more specialized journals. Two areas that have received much attention are animal cognition and applied ethology, and I shall give a few examples of such studies below. Studies on general causal models of behavior have also continued to be done, and I conclude this chapter with an examination of two such types.

ANIMAL COGNITION

The concept of cognition was long eschewed by most behavioral scientists because to them the concept implied the very sort of anthropomorphism, subjectiveness, and teleological thinking that Tinbergen praised Lorenz for removing from ethology. Over the past 30 years, however, cognition has returned to scientific respectability among psychologists, and, more recently, among animal behaviorists. In 1992, Griffin published the book *Animal Minds*, which led many ethologists to reconsider what the concept of cognition might mean. More recently, Shettleworth (1998) has reviewed the many facets of contemporary work on animal cognition.

Although there is still no general agreement about the meaning of the word, I find a simple definition quite adequate: cognition means knowing (Hogan, 1994). With this definition, facts, ideas, memories, intentions, and the like become cognitive structures, basically specific types of perceptual and central mechanisms that the animal possesses. The study of cognition in animals thus begins with asking what kinds of cognitive structures each species or individual possesses; one can then continue by asking how these structures develop, how they are activated, what function they serve, and how they have evolved.

As might be expected, these questions are often asked with respect to primates, because many people are interested in comparing the cognitive abilities of humans with non-humans, and primates are generally considered to be the group of species most closely related to humans. Many of the earlier studies were based on field work and were usually concerned with functional questions (e.g., Cheney and Seyfarth, 1990). More recently, laboratory studies of causal questions have been carried out, frequently with respect to whether or not other primates understand the behavior of other animate beings, i.e., whether they have a "theory of mind" (Povinelli *et al.*, 2000; Tomasello, 2000). A review of these studies is beyond the purview of this chapter, but I will mention two sets of experimental work on non-primates that illustrate the range and scope of causal studies of cognitive issues in animal behavior.

My first example concerns the food-caching behavior of scrub jays (*Aphelocoma* spp.). These birds scatter hoard food; they place their gathered food items in various places in their local environment and cover them with substrate. Later they retrieve and consume the stored food. Many studies have shown that scrub jays (and many other food-caching species as well) remember what they have stored and where it is located (for review see Sherry, 1985), but Clayton and Dickinson (1998) and their collaborators are currently investigating whether the birds also remember when they stored their food. This is an important question because many psychologists who study human memory have suggested that only humans are capable of "episodic" memory, a type of memory that requires knowledge of when something happened, as well as what and where it happened (Tulving, 1983; Clayton *et al.*, 2003, 2007).

Scrub jays have proved to be an excellent species to study this question because they hide both perishable and non-perishable food items; they prefer the perishable items, but have been shown to search for the non-perishable items when the elapsed time between storing and recovery is long enough for the perishable items to have decayed. However, this result could be due to differential forgetting of the caches, and not due to an explicit representation of the time of caching in memory. Recently, Selvino de Kort (de Kort *et al.*, 2005) has explored this issue in a series of experiments using perishable waxworms and non-perishable peanuts as food items. De Kort trained one group of scrub jays to expect fresh waxworms after a short cache-recovery interval and decayed waxworms after a long interval (normal group); another group of jays was trained to expect decayed waxworms after a short interval, but fresh waxworms after a long interval (reverse group). The two groups showed contrasting patterns of search after short and long cache-recovery intervals that reflected their experience: the normal group searched for waxworms after a short interval, but peanuts after a long interval, while the reverse group showed the opposite pattern. Thus, these experiments provide strong evidence that the birds not only remembered what and where they had hidden these items, but also that they had an explicit memory of when they had done it.

Clayton and her colleagues have continued exploring issues concerning the cognitive abilities of scrub jays. In a recent set of studies, they found that the jays would preferentially cache food in a place where they have learned that they will be hungry the following morning; the birds also differentially cached particular foods according to what their preferences would be the following morning. These results thus provide experimental evidence indicating that "jays can spontaneously plan for tomorrow without reference to their current motivational state, thereby challenging the idea that this is a uniquely human ability." (Raby *et al.*, 2007, p. 919.) As Shettleworth (2007) notes, studies such as these make it possible to tackle general questions of cognitive evolution (see also Byrne, 2007).

My second example concerns the behavior of jumping spiders that prey on other spiders. Robert Jackson and his collaborators have been studying for several years the prey-catching behavior of *Portia fimbriata*, an araneophagic salticid spider found in northern Australia. Based on observations and extensive experimental evidence, they have shown that *Portia* uses different tactics for catching its prey, depending on the species of prey (Jackson and Wilcox, 1998). Of particular interest for the question of cognitive abilities, they have shown that for novel prey species, *Portia* chooses its hunting techniques on the basis of trial and error learning. When hunting web-spinning spiders, *Portia* uses aggressive mimicry: "after entering the other spider's web, *Portia* sends web signals across the silk, and sometimes the resident spider responds as if a small insect were ensnared in its web. When the duped spider approaches, however, *Portia* lunges out and catches it." The signals that *Portia* sends vary according to the species being preyed upon, and are fine tuned in relation to the behavior of the individual prey. Other types of tactics are also used. In a recent series of experiments, Jackson *et al.* (2002) demonstrated that *Portia* exploits opportunistically the times when its prey's (*Zosis genicularis*) attention is focused on its own prey. Their experiments provide evidence that the responsiveness of *Zosis* to cues from potential predators is reduced when it is busy wrapping up prey, and that *Portia*, relying on optical cues alone, moves primarily during those intervals. Other studies have shown similarly complex factors influencing the hunting behavior of other species of salticids (Li *et al.*, 2003).

These two examples of causal studies of memory, planning, and decision making show that it is possible to study cognitive issues in species that most people would presume have little or no cognitive potential, using experimental methods that avoid any hint of anthropomorphism or subjectiveness. As with neuroethology and behavioral endocrinology, animal cognition has become a field in its own right, with its own journals, conferences, and textbooks. Nonetheless, it retains many of the same concepts and methods that have traditionally been used in ethology, and tends to study the organization and function of the brain at the whole-animal level of analysis.

APPLIED ETHOLOGY

One area in which studies of causation have been particularly numerous is in the field of applied ethology. Most of these studies seek answers to practical problems that confront farmers and others who raise animals commercially or people who keep animals as pets. An example of such a problem is the occurrence of feather pecking in commercial flocks of poultry. This behavior consists of pecking directed toward the plumage of other birds that often results in a feather being plucked and eaten. Feather pecking is economically undesirable because it results in decreased egg production, increased food intake, and increased mortality; it is ethically undesirable because it causes pain and may lead to skin lesions and cannibalism. Dozens of studies have been done seeking the cause of this behavior, but, to date, no management strategies have succeeded in preventing its occurrence.

There have been two main theories of the cause of feather pecking: it has been considered to be redirected ground pecking controlled by the foraging system (Blokhuis and Arkes, 1984; Blokhuis, 1989) or to be redirected pecking controlled by the dustbathing system (Vestergaard and Lisborg, 1993; Vestergaard, 1994). In both cases, deficiencies in some aspect of the environment during early life were presumed to be responsible for the development of abnormal substrate preferences for pecking. Two large-scale studies have investigated the influence of early rearing conditions on the development of feather pecking (Johnsen et al., 1998; Nicol et al., 2001). In both studies, chicks were reared either on wire mesh or on a sand or wood shavings substrate for varying periods of time; they were later tested on the various substrates. Although there were many differences between the two studies, the general conclusions were the same: All the factors that were manipulated had significant effects on feather pecking and dustbathing. Early rearing on wire mesh resulted in the most feather pecking, but rearing on more normal substrates did not eliminate the behavior. Further, regardless of the early rearing substrate, the current substrate strongly influenced all the behaviors measured. Thus, both theories received some support, but it is clear that feather pecking is not caused by any single factor.

Recently, Riedstra and Groothuis (2002) have proposed that feather pecking develops out of social exploration (see also Chow and Hogan, 2005). They confronted chicks with familiar or unfamiliar peers and found that feather pecking was higher toward the unfamiliar birds. This result is consistent with the hypothesis that stress mediates the expression of feather pecking. It is likely that foraging and dustbathing on wire mesh are also stressful, so it may be that eliminating stress will eliminate feather pecking. Unfortunately, the conditions under which commercial flocks of birds are kept are inherently stressful, which suggests that it is probably impossible to eliminate feather pecking. Nevertheless, the results of many of these studies do suggest ways of reducing the impact of the behavior.

A theoretical aspect of applied ethology relates to the problem of animal welfare. As with cognition, there is no universally accepted definition of welfare. It is generally agreed that good health or physical well being is a component of welfare. What is more controversial is to what degree psychological well being is also a component: do animals have "ethological needs" (Hughes and Duncan, 1988; Dawkins, 1990)? This concept implies that the welfare of animals will suffer if they are unable to express a normal range of behavior patterns. The notion that expressing normal behavior is pleasurable can be found in the scientific literature at least as far back as William James (1890). James wrote: "To the broody hen the notion would probably seem monstrous that there should be a creature in the world to whom a nestful of eggs was not the utterly fascinating and precious and never-to-be-too-much-sat-upon object which it is to her" (vol. 2, p. 387). He suggested that "every creature *likes* its own ways." Lorenz (1937) also proposed that performing fixed action patterns is pleasurable.

Although pleasure is a subjective concept, it is possible to define it in an experimentally meaningful way. One way to do this is to say that performance of an action or perception of a stimulus is subjectively pleasurable if an animal actively seeks out a situation in which the action is released or the stimulus is present. In practice, it is possible to demonstrate this by providing a releasing situation as a reinforcer in a standard learning task and showing that the animal learns (Hogan, 1990; Duncan, 2002). There is now considerable evidence that animals will learn various tasks when the result is allowing the animal to perform its normal behavior patterns. Rodents will learn to press a lever or run through an alleyway when allowed to dig in the sand or retrieve a pup or hoard a food pellet. Some species of song birds will learn to sit on a perch that allows them to hear conspecific song. Siamese fighting fish (*Betta splendens*) will learn to swim through a tunnel when allowed to fight with a conspecific (for a review see Hogan and Roper, 1978). It could be argued that some outcomes that an animal works for are not hedonically positive, such as a human looking up a telephone number, or, more generally, seeking information; even if the immediate outcome of such behavior is not pleasurable, it is almost certainly hedonically preferable to not receiving the information. The question remains, however, whether animals suffer if they are not allowed to engage in these activities. Several recent studies have been designed to investigate this question.

Dustbathing is one of the major behavior systems of fowl. The behavior consists of a sequence of coordinated movements of the wings, feet, head, and body that serve to spread dust through the feathers. It occurs regularly, and bouts of dustbathing last about half an hour in adult fowl. When dust is available, dustbathing functions to remove excess lipids from the feathers and to maintain good feather condition. A number of experiments have shown that the tendency to dustbathe increases as a function of the time since last dustbathing (see below). Widowski and Duncan (2000) asked whether hens that were deprived of dustbathing would work harder to obtain access to a dustbathing substrate than hens that were not deprived. One would predict that, if hens suffer when deprived of the opportunity to dustbathe, they would indeed work harder for dust access. The task required in this case was pushing open a door that gave access to a container of peat moss, a highly desirable substrate for dustbathing; the door could be made more difficult to open by adding weights. Their results were somewhat equivocal in that while most hens worked harder when deprived, a few actually worked harder when not deprived. Nonetheless, when dust access was achieved, all hens showed much more dustbathing when deprived than when not deprived. The authors conclude that the results strongly support the hypothesis that performance of dustbathing leads to a state of pleasure, though they only weakly support the notion that the hens suffer when not allowed to dustbathe.

In another set of studies, on social relationships in horses, van Dierendonck (2006, pp. 142–145) asked whether allogrooming and social play in foals should be considered "ethological needs." Her results show that these behaviors do meet all the criteria for ethological needs that have been proposed, and therefore opportunities to engage in these behaviors must be provided to ensure adequate welfare. In general, studies on questions of welfare are becoming much more numerous, and in many cases are leading to improved housing and general maintenance for many species of domestic and laboratory animals (e.g., Würbel, 2001; Wolfer et al., 2004).

## CAUSAL MODELS

Although studies of causation in the ethological literature have become much less numerous since Tinbergen's paper was written, some workers have continued to ask causal questions. One theme that has been especially productive is the development of causal models of behavior. Some of these models have been quite specific to particular behavior systems (e.g., feeding: Booth, 1978), while others have been more general and are often based on control theory concepts (e.g., McFarland, 1974, 1989; Toates, 1986; Toates and Jensen, 1991). Two types of models have been developed that can address a wide variety of motivational issues: artificial neural network models and traditional models. I give some examples of both.

### Neural network models

The use of artificial neural networks for understanding human cognitive processes began in earnest in the 1980s (Rumelhart and McClelland,

1985; McClelland and Rumelhart, 1985). Use of these models spread quickly to other fields including ethology. As Enquist and Arak (1993, p. 446) point out: "Even the most simple artificial networks, consisting of a few interconnected cells, exhibit many of the properties shown by animal recognition systems: they are easily trained to classify objects and perform generalizations. Such networks provide convenient tools for uncovering general principles of recognition free from much of the complexity found in the nervous systems of real organisms." They used such a model to show that the mechanisms concerned with signal recognition have inherent biases in response, and that these biases can act as important agents of selection on signal form. Their demonstration used a simple neural network that was trained to discriminate long-tailed birds (conspecifics) from short-tailed birds. After training, the model was tested with a wide array of stimuli and was found to respond more strongly to a stimulus with a longer tail than to the stimulus with which it was trained. If we assume that the recognition mechanism possessed by a female of the species has similar properties, and that this mechanism is attached to her sexual central mechanism, we see that she would prefer a male with an exaggerated trait; this could provide the basis for selection of the signal form. Although this demonstration was originally met with some skepticism (Dawkins and Guilford, 1995), further work has confirmed the usefulness of these models (Ghirlanda and Enquist, 1998; Enquist and Ghirlanda, 2005). In particular, Ghirlanda and Enquist (1999) and Ghirlanda (2002) have used artificial neural networks to understand the many phenomena associated with stimulus generalization. An amusing example is provided by an experiment in which chickens were required to discriminate between human faces. In the test, they were found to prefer the same faces that human subjects rated as more beautiful (Ghirlanda et al., 2002). Other workers have used these models to test proposals of how evolution might have occurred in specific groups of animals (Ryan, 1998; this volume).

It is worth noting that some of the models of the early ethologists are really very simple neural networks. The most fully developed models are those of Baerends and his colleagues (Baerends, 1976; Baerends and Drent, 1982). Baerends' model of the releasing mechanism (Baerends and Kruijt, 1973, p. 25), for example, has the same three layers characteristic of artificial neural networks: it has receptor units that respond directly to the stimulus, evaluation units that function in much the same way as the hidden units in neural networks, and a summation unit that controls the output to motor mechanisms. The major difference between Baerends' model and most neural network models is that the weightings of the evaluation units do not change as a function of their output. In other words, most neural network models incorporate the possibility of change as a function of experience, whereas Baerends' model is static in that respect. Baerends' model can be considered an expanded version of one of the perceptual mechanisms depicted in Fig. 3.1. In fact, the general type of structure depicted in Fig. 3.1 is one form of an artificial neural network.

### Traditional models

Under this heading, I include models that provide theories or formal statements of how behavior systems are believed to work. I will give examples of two such models, one for the dustbathing system in fowl and the other for the sleep system in humans. I will conclude this section with a discussion of how such models can be generalized to provide a general framework for studying causation.

As we have seen above, dustbathing is an important behavior for fowl (and many other species as well). Vestergaard (1982) gave adult fowl limited access to dust and found that the latency to dustbathe was inversely related, and duration of bouts was directly related, to the length of dust deprivation. Similar deprivation experiments on young chicks led to similar results (Hogan et al., 1991). One way to explain these results is to postulate the build-up of endogenous motivation during the period of deprivation. Most authors, however, have tried to find an explanation in terms of changes in the external stimulus situation (see Hogan, 1997). For example, the density or condition of lipids in the feathers could regulate the amount of dustbathing. It is also possible that there is a build-up of ectoparasites or other changes in skin condition during deprivation. These hypotheses were tested in experiments by several different investigators, none of whom was able to provide convincing evidence supporting the view that increased dustbathing was primarily due to peripheral factors.

Perhaps the clearest evidence demonstrating the important role of endogenous motivation comes from some recent experiments using genetically featherless chicks (Vestergaard et al., 1999). These chicks dustbathe in the normal way and with normal frequency even though they have no feathers to clean. In the experiments, chicks were deprived of the opportunity to dustbathe for one, two, or four days, and then allowed access to dust for one hour. Prior to access, the experimenters gently cleaned the chicks with potato flour to ensure that their skin condition was similar at all deprivation lengths. The results were essentially the same as the results from normal feathered chicks: The longer the period of deprivation, the more a chick dustbathed. Two-week-old chicks dustbathed for about 15 minutes after one day of deprivation, for about 20 minutes after two days of deprivation, and for about 30 minutes after four days of deprivation. Vestergaard noted that these results provide one of the best examples of a Lorenzian endogenous process governing behavior. He also suggested that his results support the existence of behavioral needs that can be satisfied only by performance of the behavior (see above).

Chickens not only dustbathe more the longer the time since they last dustbathed, they also dustbathe more in the middle of the day than at other times. Vestergaard (1982) suggested an influence of a circadian clock on dustbathing, and subsequent experiments confirmed these results (Hogan and van Boxel, 1993). More recent experiments (Hogan, unpublished data) demonstrate conclusively that a circadian clock is indeed a causal factor for dustbathing. Hogan and van Boxel (1993) proposed a two-factor model of dustbathing: one factor varies as a function of time since the occurrence of dustbathing and the second is a threshold that varies as a function of time of day. Recently, Hogan and Ghirlanda (2003) have devised a simple mathematical formalization of this model which accounts well with the results of previously published studies as well as the results of new experiments designed to test the model.

A second example is the two-process model of sleep originally proposed by Borbély (1982) and quantified by Daan et al. (1984). The first process, called the homeostatic process, increases during wakefulness and decreases during sleep. The second process is the sleep threshold, which varies as a function of time of day and is presumed to be controlled by a circadian pacemaker. This model was formulated to understand the normal sleep–wake cycle in humans, as well as disturbances in the cycle due to such variables as shift work, jet lag, and aging. The model has stimulated large numbers of experiments and continues to provide a framework for understanding sleep regulation (Borbély and Achermann, 1999; Dijk and Lockley, 2002). Experiments by Derk-Jan Dijk and his colleagues (Dijk et al., 1992; Czeisler et al., 1994) show how this model can be used to understand the relatively stable level of alertness and performance characteristic of most people during the waking day. These experiments made use of a technique called the forced desynchrony protocol. It is possible to uncouple the circadian cycle (measured by a subject's temperature) from the sleep–wake cycle if the subject is kept under conditions of low light intensity and told when to sleep and when to wake. Using this technique, subjects in these experiments were placed on a 28-hour sleep–wake cycle, in which they slept for one-third of the time and were kept awake for the other two-thirds. Their subjective alertness and their performance on simple math problems were measured. Because the subjects' circadian cycle (about 24 hours) and sleep cycle (28 hours) were uncoupled, it was possible to test their performance at different times of the circadian cycle when they had been awake for varying lengths of time. The results showed that subjective alertness was highest just after waking and declined monotonically as a function of the length of time they were awake. With respect to the effects of the circadian cycle on alertness, subjects were least alert at the part of the cycle that coincides with normal waking time, but became more alert through the course of the day until there was a sharp drop in alertness at the part of the cycle corresponding to normal sleep time. When these two results were combined (as suggested by the two-process model of sleep), subjective alertness was high in the morning and remained high until the end of the normal waking day, which is the actual result from subjects tested under normal conditions. The results further suggested that the two processes each made an equal contribution to alertness.

In many ways, the occurrence of both dustbathing in fowl and sleep in humans can be described by a model similar in principle to the original Lorenz (1937; 1950) psycho-hydraulic model of motivation (Fig. 3.3). In this model, the level of fluid in the reservoir represents the "energy" available, or the "behavioral state of readiness" for performing a specific behavior pattern such as dustbathing or sleep. The reservoir itself together with the trough represents the structure of the motor mechanism, while the spring valve, which prevents the continuous outflow of fluid, can be considered to represent the threshold. Lorenz originally proposed that the energy available increased as a function of time due to endogenous neural activity, and decreased when the behavior was actually performed. This formulation can account for some of the results for behaviors such as dustbathing, sleep, and feeding, but is not so successful with other behavior systems such as sex and aggression. However, if one allows other factors such as the priming effects of stimuli and various chemicals to contribute to the level of activation, and factors such as performance of other behavior (displacement activities) and passage of time (a leaky reservoir) to lower the activation, then many more results can be accounted for (Hogan, 1997).

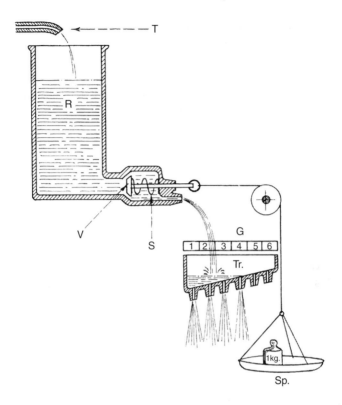

Fig. 3.3. Lorenz's model of motivation. T is the tap supplying a constant flow of endogenous energy to the reservoir, R; the valve, V, represents the releasing mechanism, and the spring, S, the inhibitory functions of the higher coordinating mechanisms; the scale pan, Sp, represents the perceptual part of the releasing mechanism, and the weight applied corresponds to the impinging stimulation; when the valve is open, energy flows out into the trough, Tr, which coordinates the pattern of muscle contractions; the intensity of the response can be read on the gauge, G. (From Lorenz, 1950.)

The generality of the model can also be increased by expanding the factors contributing to the threshold. In the original model, the fluid is held in check by the releasing mechanism, which is activated by appropriate external stimuli. One can imagine that the strength of the spring is also influenced by factors such as specific stimuli, the time of day (circadian clock), and inhibition from other behavior systems. For example, specific stimuli such as increased light intensity make behaviors such as dustbathing or sleep more or less likely, respectively. Further, the circadian clock has been shown to have a very strong influence on sleep and feeding in most species, as well as on dustbathing in fowl and parenting in female ring doves (Silver, 1990). Hogan (1997) has discussed how a two-factor model of motivation (energy or activation and threshold) can be used as a framework for studying motivation in a way that is similar to the use of the behavior system model (Fig. 3.1) as a framework for studying structure.

SUMMARY AND CONCLUSIONS

In the 40 years since Tinbergen (1963) published his views on aims and methods of ethology, ethology has ceased to exist as a unified field. In the contemporary and much broader field of behavioral biology, Tinbergen's four questions continue to be asked, but few workers attempt the sort of integration that Tinbergen's views imply. The question of causation or mechanism has frequently been addressed at more molecular levels of analysis, but a major purpose of this essay has been to show that causal questions are still relevant at the level of behavior systems in the individual animal. For this purpose, level-adequate concepts (von Holst and von St. Paul, 1960) are necessary, as illustrated by the general model of motivation described above. In this model, structure (threshold) and activation (energy) are the important concepts. They are distinguished because of their usefulness in thinking about causal factors. For example, even with the highest values for all the structural factors (best time of day, supernormal releasing stimulus, no inhibition from other systems, etc.), no behavior will occur if factors contributing to the activation variable are absent. On the other hand, even with low values of all the structural factors, behavior will occur if the activation factors are sufficiently strong (e.g., as a vacuum activity). It is necessary, however, to realize that structure and activation are relative concepts. Mathematically, these two variables are actually interchangeable (Nelson, 1965). Nonetheless, following Tinbergen (1951, p. 123), these two variables represent different functions at the behavioral level, which is sufficient justification for maintaining the distinction. At the neurophysiological level, this conception may no longer be useful because it is physically impossible to separate structure from energy (Lashley, 1938; Hebb, 1949). This is a clear example of behavioral concepts not being isomorphic with the neurophysiological mechanisms underlying them. A complete understanding of behavior requires not only answers to all of Tinbergen's questions, but also analysis at different levels. In this chapter I have stressed the importance of analysis at the level of behavior systems in the individual animal, which is in fact the level studied by the original ethologists.

ACKNOWLEDGMENTS

Preparation of this paper and some of the work described herein was supported by a grant from the Natural Sciences and Engineering Research Council of Canada.

## 4

# Tinbergen's fourth question, ontogeny: sexual and individual differentiation

Ontogeny of behavior, the development of behavior within the individual, was the fourth pillar of Tinbergen's 1963 treatise that marks the new synthesis of modern ethology. It was his contribution to Huxley's (1914, 1923) "three major problems of biology," namely causation, survival value, and evolution. Here he recognized that, while development had long been a subject of study by embryologists, for behaviorists it had become "a field in which there is a real clash of opinion." The controversy reflected the different approaches practiced by students of behavior in Western Europe and North America.

The European approach was strongly influenced by Lorenz, one of the founding fathers of Ethology. Lorenz is perhaps best known for his discovery of sexual imprinting, namely that the young of precocial birds would, if presented during a certain window of time following hatching, follow any conspicuous stimuli as if it was their mother, despite the fact that the stimulus did not have the slightest similarity with a conspecific, and that this attachment even transformed in adulthood to sexual preferences for that stimulus. This, and the development of species-specific motor patterns, led him to the proposition that the development of certain behaviors is pre-programmed by genes and independent of experience. This view was rejected vigorously by behavioral scientists in North America, most notably Hebb (1953), Lehrman (1953), and Beach and Jaynes (1954). Lehrman's "Critique of Konrad Lorenz's Theory of Instinctive Behavior" was particularly influential, in part because of the journal in which it appeared, the recognized skills of the author in the observation of animal behavior, and the personalized nature of the writing. In this view the subtle interplay of internal and external processes were considered essential in the development of any behavior, including imprinting. Such diametrically opposed opinions led to Tinbergen's understatement that "there

*Tinbergen's Legacy: Function and Mechanism in Behavioral Biology*, Johan J. Bolhuis and Simon Verhulst (eds.). Published by Cambridge University Press. © Cambridge University Press 2008.

is no agreement about the nature of the problems involved, and while the methods applied by psychologists and ethologists begin to resemble each other so closely as (in some instances) to be indistinguishable, the interpretation of the results gives rise to much discussion." (Tinbergen, 1963, p. 423).

Tinbergen recognized this dichotomy as "heuristically harmful." Indeed, the nature of scientific discovery, based as it is on cumulative knowledge, means that preconceptions often poison perception, in this case leading to intransigence. Faced with the delicate situation of writing a paper in honor of his friend's 60th birthday and, at the same time, disagreeing with Lorenz's perspective of behavioral development, Tinbergen sought to elevate the debate and proposed a new paradigm for further research. "If I were to elaborate this further I should have to cross swords with my friend Konrad Lorenz himself ... but this is not the occasion to indulge in swordplay, and I prefer to continue with my sketch of the procedure which seems to me more fruitful." (Tinbergen, 1963, p. 425).

"I believe that this discussion has been and is still being bedeviled by semantics, and that it would be helpful if, instead of discussing the justification of the use of words such as "innate" and "acquired," of "instinct" (and "instinctive") and "learning," we could return to a statement of the phenomena to be understood and the questions to be asked – indeed I think this is imperative." (Tinbergen, 1963, pp. 423–424).

Tinbergen (1963) proposed that the suitable subject of study of behavioral ontogeny should be the "change in behaviour machinery during development." This meant that "the phenomenon (change in behaviour machinery) has to be described; the problem is, how are these changes controlled? As a first step one distinguishes between influences outside the animal and internal influences." Unfortunately, like others before and after him, Tinbergen's solution did not quell the nature/nurture debate, perhaps because of the human tendency to succumb to the seduction of simplicity and dichotomies in lieu of the reality of complexity.

In real life, genes, other internal factors, and external factors continuously interact with each other. And although differences in behavior between individuals can be attributed to differences in the contribution of one of these factors, each factor is important and indispensable for the development of individual behavior. This proposition has now been firmly established in the forefront of ontogenetic research, despite ongoing discussions and misconceptions that still appear both in scientific and especially popular papers. We will discuss some examples of behavioral development within this integrative view. In keeping with Tinbergen's suggestion, we will not indulge in semantics, but try to illustrate how genetic, ecological, physiological, and social processes, and the integration among these, shape the development of social behavior.

The way Tinbergen formulated the study of behavioral development suggests it to be the study of causal mechanisms. However, young organisms have to be adapted to the environment in which they are born, and because they are different from adult conspecifics they need their own adaptations. Hence, behavioral development consists of so-called ontogenetic adaptations, and is a product of evolutionary history. In fact, all of the other of Tinbergen's "Why" questions detailed in the 1963 paper, namely description, causation, function, and evolution, can and should be applied to the study of behavioral development. It is our view that in behavioral biology questions about the causation of development (changes in the underlying machinery of behavior as Tinbergen called it) should be inspired by considerations about function: what are the problems that have to be solved by developing organisms to survive and reach maturity? We will address functional aspects of development throughout this chapter.

At the time Tinbergen wrote his paper, ethologists were mainly concerned with the development of species-typical behaviors. Differences in behavior between individuals of the same species were mostly interpreted as noise around an adaptive mean. Nowadays it has become clear that behavioral differences between individuals of the same species and sex may reveal important biological information. Therefore, we will describe some mechanisms by which differences in the social behavior between males and females as well as that of individuals of the same sex, can develop.

Mothers can influence the development of their offspring not only by genomic, but also by non-genomic effects. The advantage of the latter is that it can be adjusted to the situation in which the female is reproducing and in which the young will be reared. Interestingly, such non-genomic effects may be carried across generations that may generate patterns that may seem to be, but in fact are not, heritable. Two clear examples of this are the effects of incubation temperature and maternally derived hormones, and these will be the center of our focus. We will discuss the influence of these factors, as well as (their interaction with) social experience, on the development of behavior during three different age classes: prenatal, early postnatal or juvenile stage, and adulthood.

We are, within the tradition of early Ethologists, convinced that the use of different animal models, other than the classical laboratory rodent, can be of great help to understand behavioral ontogeny. Therefore, we will discuss examples of a variety of animal taxa, such as reptiles, birds, and rodents.

### THE EMBRYONIC ENVIRONMENT

Although the effect of developmental processes on behavior often only becomes apparent after birth or hatching, the pre-natal stage is very important for the developmental process itself. During this early stage physiological factors affect development of brain and behavior, often during a sensitive phase and with irreversible effects. Such early processes can affect a whole array of behaviors. Clear examples of this are the effects of steroid hormones on sexual differentiation, affecting sexual and agonistic behavior, and the effect of light on brain lateralization in avian species (Rogers and Deng, 2002), affecting both social and non-social behavior. Therefore most of our chapter deals with this early phase in development. We will discuss two examples in some detail, the effect of incubation temperature on sexual determination and differentiation in the leopard gecko, and the effect of prenatal exposure to maternal androgens in birds. Both examples illustrate, each in their own way, the importance of the mother in shaping individual differences in behavior in the offspring.

### Sexual determination and differentiation

*Temperature dependent sex determination in the leopard gecko as a model*

Mammals and birds have sex chromosomes. In other words, genetic sex and gonadal sex are inextricably linked. This genetic difference facilitates the study of sex differences and, indeed, one of the great success stories in genetics of the last century was the discovery of a gene on the short arm of the Y chromosome dictating whether the primordial gonadal ridge would develop into an ovary or a testis. In fact the process of sexual differentiation is very illustrative of how genes are translated to behavior via the production of gonads, that determine in turn the hormonal milieu, which in turns determines adult sexual behavior, a process with a lot of plasticity despite the strong influence of genetic factors.

But the very nature of genotypic sex determination makes it difficult to distinguish epigenetic from genetic contributions to sexuality. Consider, for example, aggressive and sexual behaviors displayed by both sexes, but at different frequencies. To what extent are differences observed between adult males and females due to their differences in sex chromosomes, differences in the nature and pattern of hormone secretion, differences in non-genomic yet heritable factors such as maternal influences, or even sex-typical experiences? So, in animals with sex chromosomes, the genetic element typically cannot be separated from the epigenetic element. This is particularly important as the sex chromosomes may be involved in the sexual differentiation of the brain as well as the gonads (Arnold, 2002).

Ideally, studies of the development of sexuality would utilize animal models that exhibit sex-typical differences in the traits of interest, yet not have the complications arising from sex-specific chromosomes. In other words, a species that can illustrate how different environments can elicit different phenotypes from a particular genotype without the confound of sex-limited genes. Do such organisms exist in nature?

Fortunately, not all vertebrates have sex chromosomes. Indeed, there are many vertebrate species in which sex is determined not by sex chromosomes, but by the environment (Crews, 1993, 2002). One form of environmental sex determination is temperature-dependent sex determination (TSD). In TSD, gonadal sex is plastic initially but becomes fixed by the temperature of the incubating egg during the mid-trimester of development (Crews, 1996). The laboratory of Crews has been studying TSD both in terms of its causal mechanisms and its functional outcomes.

We have taken advantage of the TSD system to investigate how events early in life can act directly on body and brain to shape adult sexuality independent of the gonad and its hormones. In this work we have used a TSD animal, the leopard gecko (*Eublepharis macularius*). This species also lacks parental care, facilitating the interpretation of results of incubation temperature per se. In the leopard gecko, high and low incubation temperatures produce only or mostly females, while intermediate incubation temperatures produce different sex ratios (Fig. 4.1). That is, 26 °C and 34 °C are female-producing incubation temperatures, whereas 30 °C produces a female-biased sex ratio, and 32.5 °C produces a male-biased sex ratio. It is important to note that incubation temperature and gonadal sex are not completely linked in TSD; in other words, this association is not as fixed as gonadal sex and genetic sex is in species with genotypic sex determination. Rather, the

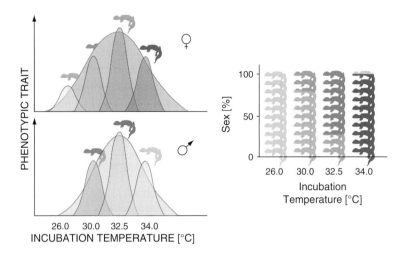

Fig. 4.1. Pattern of temperature-dependent sex determination in the leopard gecko. Right panel indicates how only females result from a 26 °C incubation temperature while both sexes are produced at 30°, 32.5°, and 34 °C incubation temperatures, albeit in different sex ratios. Left panels indicate how within each sex (females top panel, males bottom panel) the morphological, physiological, neural, and behavioral traits measured to date are different according to incubation temperature, yet when combined across temperatures, are normally distributed.

effect of incubation temperature and gonadal sex can be dissociated. Thus, the effect of gonadal sex can be determined by comparing males and females from the same incubation temperature and the effects of embryonic temperature can be determined by comparing males or females from the different temperatures. In other words, when sex differences are the focus, we compare males and females from within a particular incubation temperature, and when the development of sexuality is the focus, we compare males (or females) from the different incubation temperatures.

### Temperature-dependent sex determination and the development of individual differences

The results of a variety of experiments indicate that incubation temperature does more than establish the gonadal sex of the individual. That is, the temperature experienced during embryogenesis accounts for much of the within-sex variation observed in the morphology, growth, endocrine physiology, and aggressive and sexual behavior of the adult

(Crews, 1999; Crews *et al.*, 1998; Rhen and Crews, 2002; Sakata and Crews, 2004). For example, males in general grow more rapidly and are larger than females. That is, males from the intermediate incubation temperatures are larger than are females from the same temperatures. However, females from the male-biased incubation temperature grow as rapidly and become as large as males from female-biased incubation temperature. The temperature experienced during embryonic development also determines the relative concentrations of sex hormones that the individuals exhibit as adults. In general, there is a log unit difference in the androgen-to-estrogen ratio between males and females. However, the endocrine physiology of the adult varies dramatically with the temperature experienced during incubation. In both sexes the androgen-to-estrogen is highest in animals from the male-biased incubation temperature.

Incubation temperature also has a major influence on the nature and frequency of the behavior displayed by the adult leopard gecko. In general, females usually respond aggressively only if attacked, whereas males will posture and then attack other males but rarely females. Females from a male-biased incubation temperature are significantly more aggressive toward males than are females from a low or female-biased incubation temperature. These same females show the male-typical pattern of offensive aggression. Similar effects of incubation temperature on aggressive behavior have been documented in males.

Courtship is a male-typical behavior. In a sexual encounter, the male will slowly approach the female, touching the substrate with his tongue or licking the air. Attractiveness is a female-typical trait and is measured by the intensity of a sexually active male's courtship behavior toward her. Females from a male-biased incubation temperature are less attractive to males than are females from lower incubation temperatures. Interestingly, attractivity in females from the high ($34\,^{\circ}$C) incubation temperature is greater than that of females from male-biased incubation temperature and not different from that of low-temperature ($26\,^{\circ}$C) females. The mating preferences of males are also influenced by incubation temperature as indicated in Y-maze tests. Males from the male-biased incubation temperature prefer females from a low incubation temperature relative to females from a high incubation temperature, while males from the female-biased temperature prefer females from a high incubation temperature relative to females from a low incubation temperature (Putz and Crews, 2006) (Fig. 4.2).

Aggressive and sexual behaviors in the leopard gecko are modulated by sex hormones and incubation temperature also influences the

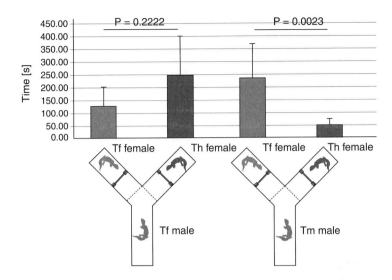

Fig. 4.2. Differential mate choice behavior in male leopard geckos from two incubation temperatures. Males from a male-biased incubation temperature (Tm) spend a significantly greater amount of time close to females from a female-biased incubation temperature (Tf) than with females from a high incubation temperature (Th), while the opposite was true for Tf males.

ability of exogenous sex hormones to maintain or restore sexual and aggressive behaviors in both sexes. For example, we find that, depending upon their incubation temperature, adult males respond differently to hormone replacement therapy following castration and, as one would predict, incubation temperature modulates patterns of brain metabolism. Similar results have been found in females. This suggests that incubation temperature influences how the individual responds to steroid hormones in adulthood.

Because the nature and pattern of growth, hormone secretion, and behavior ultimately are expressions of brain activity, it stands to reason that neural phenotypes must also exist and these might be sensitive to incubation temperature. How might incubation temperature during embryogenesis affect the brain of adult animals?

### A unitary neuroanatomical framework

Given the wealth of evidence that in animals with genotypic sex determination certain limbic nuclei like the preoptic area and the ventromedial hypothalamus are sexually dimorphic, we expected to find similar sex differences in the leopard gecko. It was a surprise to discover that there are no statistically significant sexual dimorphisms in either of these brain areas between males and females at those incubation temperatures that produce both sexes. There are, however, consistent differences across incubation temperatures. For example, the volume of the preoptic area is larger in both males and females from the male-biased incubation temperature compared to animals from the female-biased incubation temperature. The opposite pattern is found for the ventromedial hypothalamus. That is, the volume of the ventromedial hypothalamus is larger in females from low incubation temperatures compared to females from the male-biased incubation temperature. Other research has shown that, if males from different incubation temperatures are castrated and given equivalent amounts of sex steroid hormones, the difference in agonistic behavior between the different types of males persist, suggesting that incubation temperature of the embryo may directly organize the development of brain nuclei independent of gonadal sex in a manner similar to that for body growth.

We have learned a great deal from studies in which brain areas have been destroyed or chemically or electrically stimulated, but there are limitations to what they can tell us about behavior–brain relationships. While there is no disagreement that complex behaviors must reflect complex patterns of neural activity involving many areas of the brain, how to visualize such coordinated patterns of brain activity has been a challenge. One approach has been functional brain mapping using methods that include, but are not restricted to, cytochrome oxidase histochemistry, immediate early gene expression, and 2-deoxyglucose uptake that correlate patterns of behavioral expression with patterns of activity in various brain nuclei.

Cytochrome oxidase is a terminal enzyme of the electron transport chain located in the inner mitochondrial membrane and catalyzes the transfer of electrons to oxygen to form water and ATP. Thus, cytochrome oxidase is a rate-limiting enzyme in oxidative phosphorylation, the major pathway in brain metabolism. Consequently, the abundance and activity of cytochrome oxidase activity in a brain area is a measure of the metabolic capacity of that brain region and reflects the metabolic history of an area. In other words, the activity of cytochrome oxidase determines the amount of ATP available in a neuron, thereby constraining the amount of activity a neuron can sustain and can serve as a marker of metabolic capacity in brain nuclei

(Gonzalez-Lima, 1992). Cytochrome oxidase histochemistry is not like 2-deoxyglucose autoradiography or c-fos immunocytochemistry, which provide information on evoked or immediate activity; instead, cytochrome oxidase reveals long-term changes in brain activity.

Sexual behavior is particularly amenable to the study of behavior–brain relationships. A large number of experiments have identified the nuclei of the limbic forebrain to be critically involved in the display of sexual behavior. Further, a variety of methods have revealed them to be interconnected, to contain sex steroid hormone receptors, and to be sexually dimorphic in their volume and synaptic organization as a consequence of the nature and frequency of sex steroid hormones secreted perinatally. Importantly, these properties appear to be evolutionarily conserved. For example, in both mammals and reptiles metabolic activity in limbic areas reflects the capacity to display sociosexual behaviors and, in turn, that differences in metabolic activity in these areas reflect individual differences in the propensity to display social behaviors (reviewed in Crews, 1992; Sakata *et al.*, 2001).

The possibility that these limbic nuclei might form an integrated neuronal network with overlapping functions that subserve all sex steroid hormone-modulated social and reproductive behaviors has been considered previously, most recently by Newman (1999). Thus, the concept that social and reproductive behaviors may "emerge from the activity of a unitary neuroanatomical framework" in the brain (Newman, pp. 252–3), and not simply the product of activity of a single brain area(s), complements more traditional approaches of mapping different behavioral functions on subnuclei in these brain areas. Further, Newman suggested a graphical representation of the data generated in studies of metabolic activity that would allow one to see how nuclei in this network may differ in different behavioral states. It is our opinion that her proposal represents a marked improvement on the traditional tabular form of data presentation that makes it difficult to detect the relationships among the various nuclei measured. While Newman proposed hypothetical representations, Crews has recently applied and expanded this method as it relates to the neural landscape of metabolic activity of the limbic nuclei associated with social behavior. This conceptualization is similar to both Waddington's developmental landscape during embryogenesis and the peaks and valleys of a fitness landscape in theoretical biology.

We have been interested in how experience might affect the neural substrates of aggressive and sexual behavior in leopard geckos. Extending Newman's concept, we have mapped which brain areas mediate experience-dependent changes in behavior. For example, we find embryonic experience influences both between sex and within sex differences in the limbic landscapes. Relative to females, males on average have greater cytochrome oxidase activity in the preoptic area (Coomber *et al.*, 1997). The complement is also true, that is, relative to males females on average have greater cytochrome oxidase activity in the ventromedial hypothalamus. But incubation temperature is an important determinant in both sexes and these differences correlate well with behaviors exhibited by animals from different incubation temperatures. For example, males from the male-biased incubation temperature are more aggressive and have greater cytochrome oxidase activity in the anterior hypothalamus, septum, and the nucleus sphericus (homolog of the medial amygdala) compared to males from the female-biased incubation temperature (Fig. 4.3). As might be expected, females from the female-biased incubation temperature have greater cytochrome oxidase activity in the ventromedial hypothalamus compared to females from the male-biased incubation temperature. As mentioned previously, there is a significant increase in aggression in females from higher incubation temperatures. Analysis of females from different incubation temperatures reveals that cytochrome oxidase activity increases in brain nuclei associated with aggressive behavior as a function of incubation temperature in a manner that parallels the differences in aggression among females from different incubation temperatures.

## Hormone-mediated maternal effects and individual differentiation

### Sibling effects in mammalian species

As we have seen above, hormones play a key role in the differentiation between the sexes. Differences in early exposure to sex steroid hormones produce permanent differences between males and females in morphology, brain, and behavior. Such permanent effects of early hormone exposure, inducing structural changes in the brain, are called organizing effects. Part of these effects involves the sensitivity of the brain and other organs to the activating effects of hormones later in life. So far, we have been discussing the effects of hormones produced by the embryo itself. However, the embryo is also exposed to steroid hormones from other sources, namely those of the mother, and in litter bearing species, of their siblings. This can have important consequences for development.

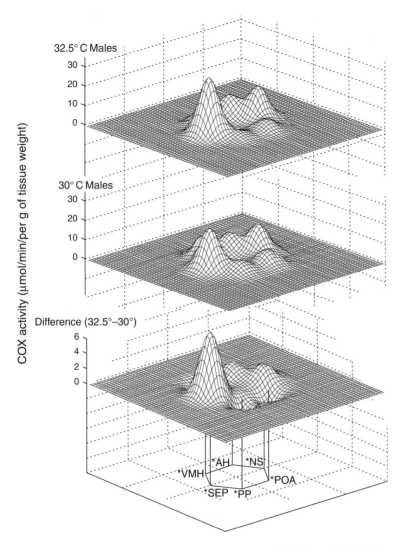

Fig. 4.3. Metabolic capacity in the interconnected limbic nuclei of adult
male leopard geckos from 32.5 °C incubation temperature (top panel)
and from 30 °C incubation temperature (middle panel). Bottom panel
indicates the difference between the neural landscapes, revealing the
effect of temperature during incubation on metabolic activity in the adult
brain. Illustrated in least squared means of cytochrome oxidase (COX)
activity. All nuclei accumulate sex steroid hormones and are involved in
the expression of social behaviors. Positive peaks indicate nuclei in which
change was more positive in males from a 32.5 °C incubation
temperature, whereas negative peaks indicate nuclei in which change
was more positive in males from a 30 °C incubation temperature. Asterisk

In litter-bearing mammals the embryos are positioned closely beside each other in the uterus. At some time during development, these embryos produce sex steroids that also reach the neighboring fetus. As a consequence, the sex of an individual's neighbors in the uterus can influence its physiology and behavior in adulthood (vom Saal *et al.*, 1999). That is, the endocrine microenvironment during fetal development modifies the physiology, behavior, and neurochemistry of the individual. Thus, a female fetus located between two males (a 2M female) is exposed to higher levels of androgen produced by the neighboring males compared to a female fetus located between two females (a 2F female). Behavioral effects of this early endocrine milieu have been demonstrated in pigs, mice, and gerbils. Largely through the work of Mertice Clark and Jeff Galef on the latter we have learned that, as adults, these 2M females have lower estrogen and higher testosterone levels, have a late onset and a long estrus cycle, have a masculinized phenotype, are less attractive to males and more aggressive to females, and produce litters with significantly more male-biased sex ratios relative to 2F females (Clark and Galef, 1995). Effects of the intrauterine position in males have also been found. In addition, Howdeshell and colleagues (1999, 2000) recently demonstrated that the fetus's intrauterine position also influences its sensitivity to exogenous hormones in adulthood. Such effects might also be relevant for explaining individual differentiation in twins in our own species.

In collaboration with Clark and Galef the laboratory of Crews has demonstrated that the metabolic activity of brain nuclei can also vary according to uterine position. In gerbils, the sexually dimorphic area of the preoptic area is responsible for copulatory behavior in males and, as females differ in their sexual behavior according to intrauterine position, this area is likely to be involved in their behavior as well. Cytochrome oxidase histochemistry reveals long-term changes in the metabolic capacity in the sexually dimorphic area of the preoptic area, with 2M females having greater activity compared to 2F females (Jones *et al.*, 1997). There also is a difference in cytochrome oxidase activity in the posterior anterior hypothalamus, an area replete with neurons

---

Caption for Fig. 4.3. (cont.)
indicates significant difference. AH (anterior hypothalamus), VMH (ventromedial hypothalamus), AME (external amygdala), SEP (septum), PP (periventricular nucleus of the preoptic area), POA (preoptic area), NS (nucleus sphericus).

containing GnRH, which may partly explain the physiological differences between 2M and 2F females.

It is clear that the intrauterine position, affecting early exposure to sex steroids, lead to differences within the same sex. The functional consequences of these differences are as yet not clear. Experiments in semi-natural environments indicate that 2M females have larger home ranges (Zielinski *et al.*, 1992), suggesting that they might do better in high density. There is some evidence that mammalian mothers can vary the sex ratio of their litters, even between the two different uterus horns ((Clark *et al.*, 1991b) and by doing so they will influence the frequency of 2M females. This opens the possibility that females can adjust the frequency of 2M females according to the situation in which they are reproducing. Whether the effects of intrauterine position are adaptive or not needs further study.

### Maternal hormonal condition during pregnancy

During pregnancy the hormone production of the mother varies, in relation to the nature and duration of the pregnancy, genetic factors, and stressful events. For example, in the Mongolian gerbil circulating concentrations of androgens in the maternal circulation is correlated with the fetal sex ratio (Clark *et al.*, 1991a), which in turn can influence the behavior of the mother. In addition, steroid hormones from maternal origin such as androgens and corticosterone and cortisol can reach the fetus via the placenta. The placenta, itself an endocrine organ, can metabolize maternal steroid hormones, thereby protecting the fetus from maternal hormones, but can do so only to a limited extent. As a consequence the hormonal status of the mother during pregnancy can affect the development of the fetus. The most conspicuous example is the case of the spotted hyena, in which females are dominant, larger and more aggressive than males and in which there is heavy sibling competition between the young pups. The clitoris of female young is greatly enlarged, forming a pseudo penis. It is thought that this masculinization and/or early aggression between pups come from the high androgen levels females have during pregnancy (e.g., Licht *et al.*, 1992; Goymann *et al.*, 2001). Similarly, gender roles of girls in humans relate to levels of testosterone in the blood of their mothers when they were pregnant of these children (Hines *et al.*, 2002).

Another well-studied example is the effect of early maternal stress. For example, handling, or treatment with stress hormones of pregnant dams, or exposing them to an unstable social environment will not only influence the behavior and physiology of the mother but also of her young. This affects their later stress sensitivity, hypothal-amo-pituitary-adrenal axis, social behavior and cognitive functions (Weinstock, 1997), as well as the masculinization of female pups (Kaiser and Sachser, 2005). Part of these effects are probably directly due to the effect of the elevated levels of stress hormone in the mother on the developing fetus, while anther part can be attributed to changes in her early parental behavior. So far, most of the effects on the off-spring are interpreted as pathological and detrimental effects. It is, however, conceivable that the Darwinian fitness of the offspring is maximized and adjusted via this maternal hormone exposure to the stressful conditions the mother is exposed to and that will be the environmental condition of the offspring too. For example, pregnant mothers that experience frequently stressful encounters with preda-tors may prepare their offspring to this situation by making them more cautious and fearful.

### Effects of maternal androgens: the avian model

Mammalian species have several disadvantages to study the effects of maternal hormones on development. First, the relationship between blood levels of hormones in the mother and what actually reaches the embryo is difficult to quantify because the placenta acts as an inter-phase between both and metabolizes hormones itself. Second, early exposure to maternal hormones fluctuates while, as we have seen, hormones from siblings can also reach the embryo. Finally, manipu-lation of exposure of the embryo to hormones, indispensable to test hypotheses of descriptive studies, is difficult since the embryo develops inside the mother's body. Fortunately, many animal species produce eggs that develop outside the mother's body. In many of such oviparous species, including fish, turtles, reptiles, and birds, these eggs contain substantial levels of maternal hormones. For two reasons birds are especially suitable for studying patterns, causes and consequences of maternal hormone deposition in eggs. First, birds produce relatively large eggs, facilitating experimentation and the determination of hor-mone levels in individual eggs. This facilitates the causal analyses of the hormone mediated developmental processes. Second, the ecology of several bird species is well known. This facilitates the analyses of functional and evolutionary aspects of these processes.

It was mainly Hubert Schwabl who opened up this field of research with the discovery that canary eggs contain androgens that affect the growth of the offspring from these eggs (Schwabl, 1993). Since then the field has rapidly expanded. So far, all bird species of which their eggs have been analyzed on steroid hormones appear to produce eggs that contain substantial levels of androgens: androstenedione, testosterone, and dihydrotestosterone. Furthermore, the levels of these androgens fluctuate systematically both within the clutch and between clutches. We will discuss both in this order.

*(i)    Within clutch variation*

Most of the attention in the literature has so far been given to within clutch variation. The majority of bird species produce clutches of several eggs that are laid with at least a time interval of one day. In several of these species levels of maternal androgens increase with the order in which the eggs of the clutch are laid. For example, in black-headed gulls (*Larus ridibundus*), androgen levels strongly increase with laying order of the three eggs of the clutch (Fig. 4.4). It has been hypothesized that such patterns are adaptive since the mother can compensate in this way for the disadvantage of the asynchrony in hatching between the different chicks (Schwabl, 1993). For example, in black-headed gulls, incubation starts already when the first egg is

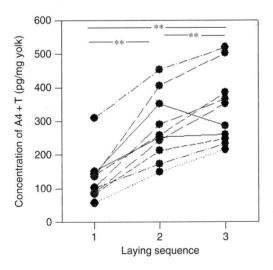

Fig. 4.4. Levels of androgens (androstenedione plus testosterone) in the yolk of freshly laid eggs of the black-headed gull in relation to the position of the egg in the laying order. Lines connect clutches. ** $P < 0.01$.

laid, so before clutch completion. As a consequence, the first egg will hatch before the second and the second before the third. This will give chicks of later laid eggs, that are younger and weaker than their older siblings, a disadvantage in sibling competition, since they completely rely on food from the parents, for which they have to compete. This increase in androgen deposition to last laid eggs may mitigate the disadvantage of hatching asynchrony since it may enhance the competitive ability of last hatched chicks. Indeed, canary chicks that hatched from eggs injected with androgens begged more frequently and grew more rapid than chicks of control eggs (Schwabl, 1996a). However, these effects were apparent only the first days after hatching and obtained under artificial conditions. Therefore the laboratory of Groothuis set out to do a similar experiment in the field. First laid eggs collected in black-headed gull colonies were found to contain relatively low levels of androgens (see above). Next, clutches of three first-laid eggs were composed and cross-fostered to other gull pairs. Each set of three eggs was selected such that the eggs would hatch about one day after each other, as in natural clutches. In experimental clutches, the egg that would hatch first was injected with vehicle only, the second with a low dose of androgens and the one that would hatch last with a higher dose of androgens. In control clutches, all three eggs were injected with vehicle. In this way experimental nests were created that mimicked the natural situation (Fig. 4.4) and "control" clutches that lacked this pattern of androgens (Fig. 4.5c). Time of hatching and growth up to the fledging stage was measured. The interval between hatching of the first and the third egg was smaller in the experimental clutches, indicating that yolk androgens shorten the degree of hatching asynchrony. Further, chicks of third eggs of experimental clutches grew faster than of third control eggs (Fig. 4.5b). Both results indicate that elevated levels of maternal androgens in the egg may indeed partly compensate for hatching asynchrony. Interestingly, chicks from first eggs of experimental clutches (HO eggs, Fig. 4.5c) that had to compete with siblings treated with hormone (H1 and H2 eggs) did not grow as fast as first hatched chicks from control clutches (OIL A) that had siblings that received vehicle only (Fig. 4.5a). This suggests that first chicks in experimental clutches suffered from increased competition with their siblings that had hatched from eggs treated with androgens. This suggests that yolk androgens indeed affect competitiveness (Eising et al., 2001).

The latter interpretation was supported by another field study. In this instance the Groothuis laboratory composed clutches of two first laid eggs, matched for weight and hatching time. One was injected with

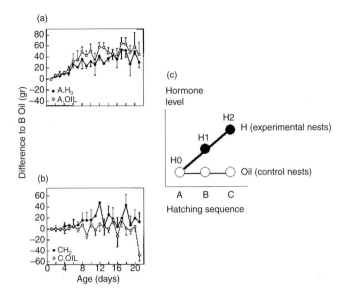

Fig. 4.5. Body mass changes with age of chicks hatching from eggs treated with different doses of a combination of androstenedione and testosterone. Plotted are the differences with body mass of chicks that hatched from B eggs in oil clutches. A: First hatched chicks of experimental clutches (hormone clutches in which eggs were treated with increasing levels of androgens: H0 = oil only, H3 is highest dose of androgens) grow less well than first hatched chicks of control nests (all eggs injected with oil only), due to increased competition by siblings of H1 and H2 eggs. C: Third hatched chicks from experimental chicks, receiving the high dose of androgens, grow better than third hatched chicks of control eggs. B: experimental design.

androgens, the other with vehicle. After hatching, the behavior of the chicks was recorded at the times the parents came in to feed the chicks. Chicks of androgen treated eggs were more alert, reacted more often as the first one to the parent, begged more, and obtained a larger share of food. This indicates that the effect of yolk androgens on growth is mediated by the effect of the androgens on behavior (Eising and Groothuis, 2003). These data fit the finding (Lipar and Ketterson, 2000) that yolk androgens enhance the development of the neck muscle in the very young chick, since this muscle is important both for begging (a conspicuous pumping movement of the neck and head) and for hatching behavior. In the same experiment early survival was enhanced in chicks from androgen treated eggs, indicating a beneficial effect of maternal androgens on an important fitness parameter.

Although these data support the hypothesis that the pattern of androgen deposition over the eggs of a clutch is adaptive, they do not allow the conclusion that individual mothers are able to adjust this allocation pattern to the actual level of hatching asynchrony that shows considerable variation among and within species: A higher degree of hatching asynchrony should require a stronger increase of hormone levels over the laying sequence than a smaller degree. The problem of disposition adjustment has been approached at the level of between-species comparison. For example, in some species that show hardly any hatching asynchrony, changes in androgens over the laying sequence are less strong. Moreover, in a siblicidal species, in which the oldest sibling, when it hatches healthy, kills off the younger one, to adjust the brood size to food availability, mothers deposit more androgens in first laid eggs perhaps to facilitate the process of siblicide (Schwabl et al., 1997). However, this does not tell us whether within species individual mothers are flexible in their deposition strategy.

The Groothuis laboratory has tried to answer this question by looking at variation in hormone allocation within the same species, again the black-headed gull. To this end, full clutches were collected in the field just after the last egg was laid. Eggs were then incubated for a standard time, after which the size of the embryo was measured and yolk hormone levels determined. Embryo size of the first laid eggs was always larger than that of the last laid egg, because it had already been incubated for some time in the field. Based on an independent data set of eggs incubated for various periods of time, the length of time the first egg had been incubated before incubation on the third egg started could be estimated on the basis of embryo size. This, in turn, gave an indication of the expected degree of hatching asynchrony. This estimate positively correlated with the increase of androgens over the laying sequence. In other words, clutches in which hatching asynchrony was expected to be large showed also large differences in androgen content between the third and first egg. The degree of hatching asynchrony is dependent on the incubation pattern. Incubation in birds is under the control of prolactin that at the same time influences androgen production. Therefore these data not only provided evidence for individual adjustment of maternal hormone allocation, but also a mechanism underlying this (Mueller et al., 2004).

However, the relation between hatching asynchrony and androgen deposition in the egg is probably more complex. Whether mitigating the effect of hatching asynchrony is adaptive depends on the function of hatching asynchrony itself, for which many hypotheses have been put forward, and on the extent to which avian mothers are able or constraint to manipulate the degree of hatching asynchrony. For example, in open ground breeders such as gulls, parents may have to sit already on their first egg when egg predation is high or the weather conditions such as sun radiation may impose detrimental effects on egg viability. Differential androgen deposition may then be the only option to counteract the by-effect of early incubation on hatching asynchrony. In addition, in species in which siblicide only occurs under poor food conditions in which parents can't rear the full brood, the androgen deposition pattern may strongly depend on the current food situation. (For a more extensive discussion of this topic see Groothuis *et al.*, 2005.)

*(ii)    Between-clutch variation*

Adjustment to hatching asynchrony cannot be the sole explanation for variation in maternal androgen allocation to the eggs, since there is also considerable variation in androgen levels between clutches. Further, all eggs, and not only those late in the laying sequence, contain these hormones (Fig. 4.4). At least part of this between clutch variation is related to the social environment of the mother. It is well known for many animal species, including birds, that the production of gonadal steroid hormones is influenced by social factors such as the presence and quality of the other sex and the level of social competition. Such factors may stimulate androgen production in the female around the time most of the yolk of the eggs are deposited, a few days before ovulation and egg laying. Hormone levels in the egg are then a reflection of the hormonal state of the mother, induced by the social situation. We have found some evidence for this. Black-headed gull pairs that have larger territories and are most likely more aggressive (aggression being under the influence of androgens) produce clutches with relatively high levels of maternal androgens (Groothuis and Schwabl, 2002). The same was found for pairs nesting in places for which it is likely that a high level of competition exists (Groothuis and Schwabl, 2002). Evidence from other studies with other bird species is consistent with this (for review see Groothuis and Schwabl, 2002). In addition Gil *et al.*, (1999) found enhanced levels of androgens in eggs of females paired with attractive mates compared to those paired with unattractive mates, a finding recently replicated in a more natural set-up (Von Engelhardt *et al.*, submitted).

The question can be raised whether the relation between the social environment of the mother and hormone levels in her egg is just an epiphenomenon or an adaptation. Given the strong effects of maternal androgens on development we believe the latter to be the case. By providing her broods with more androgens when the mother experiences a higher level of competition she may prepare her offspring for a higher level of competition they may encounter after hatching. By doing the same when paired with an attractive male, she may invest more in offspring from a high quality father (assuming that higher deposition requires higher circulating levels of androgens in the female that may impose some costs to her, cf. Clotfelter *et al.*, 2004). However, the first explanation cannot apply for altricial species in which brood competition only occurs after fledging. This suggests that exposure to maternal hormones in the egg may have long-term consequences for the phenotype of the offspring, either because it adjusts the phenotype to the level of competition that the individual may encounter during reproduction in the natal colony, or because the production of different phenotypes in itself is advantageous.

*(iii)    Long-lasting effects*

We have seen that exposure to elevated levels of maternal androgens in the egg can enhance growth and thereby fledging weight (Eising *et al.*, 2001). It has been demonstrated in several bird species that fledging weight determines survival in first winter, and also later social status and reproductive success. In this way maternal androgens can have long-lasting effects on the offspring. In addition, these hormones may lead to individual differentiation in a more direct way. It is well known that early exposure to androgens can have organizing effects on brain and behavior. Indeed, manipulation of androgen and estrogen levels in bird eggs is well known to affect sexual differentiation. However, by manipulating levels of androgens in the eggs within the range of natural within-clutch differences, we applied levels that were much lower than those used in studies on sexual differentiation. Still, these manipulated levels affected begging behavior up to the age of 4 weeks (Eising and Groothuis, 2003). Moreover, it was found in house sparrows that androgen levels in the yolk correlated positively with social rank order among juvenile birds hatched from these eggs (Schwabl, 1996b). However, this was a correlational study and since yolk levels of androgens can correlate with other aspects of the egg

(Groothuis and Schwabl, 2002), an experimental approach is required to answer this question. To this end, data on social behavior and nuptial plumage of black-headed gulls were collected ten months after they had hatched from either eggs injected with androgens or vehicle only. The data provide strong evidence for long lasting effects of maternal hormones (Eising *et al.*, 2006). This suggests that by providing her broods with more androgens in areas of high competition, the mother translates environmental conditions, affecting her own hormone production, to the next generation.

### EXPERIENCES DURING INFANCY AND IN ADULTHOOD

So far, we have emphasized the important role of pre-natal environmental factors on the development of social behavior. However, this is not the only important life history phase for the development of this behavior. After hatching or being born, the animal starts to interact with its social environment. Numerous studies have elucidated the role of social experience in shaping social behavior. We will illustrate the influence of some specific social factors in leopard geckos, mice, and birds respectively.

### Social experience in the leopard gecko

Experience during adulthood can also influence brain neurochemistry and behavior in leopard geckos. It is well known that, relative to sexually naïve males, sexually experienced male rats and cats initiate copulation sooner, tend to be more aggressive, continue to copulate longer after castration, and respond more rapidly to androgen replacement. It is also common to find in mammals that experienced males exhibit greater changes in sex steroid hormone concentrations and immediate early gene expression when presented with cues that predict the introduction of a female. Altogether, it appears that sexually experienced males are more primed for sexual behavior.

Similar effects of sexual experience are found in the leopard gecko. For example, sexually experienced male geckos begin to mark sooner, are less likely to flee from a territorial male, and have higher circulating concentrations of testosterone than naïve males (Crews *et al.*, 1998; Sakata *et al.*, 2002). In general, we find that sociosexual experience also increases metabolic capacity in certain nuclei but reduces it in others; there also are nuclei where there is no discernible effect. But this effect of sexual experience on both volume and metabolic capacity of brain nuclei is dependent upon incubation temperature. Again, the effects can vary from brain area to brain area. For example, the volume of the preoptic area increases with sexual experience in low-temperature females, but not in females from the male-biased incubation temperature, whereas cytochrome oxidase activity in the ventromedial hypothalamus increases in females from the male-biased incubation temperature, but not in low-temperature females.

Another wrinkle comes in when we consider age. We have looked at this question by incubating eggs at different temperatures, raising hatchlings in isolation before transferring them to breeding cages on their first birthday or several years later. In leopard gecko life this corresponds to an 18-year-old human with a 36-year-old. In this way we are able to assess the relative effects of age independent of social and sexual experience and further, to determine if embryonic experience could affect the response.

As mentioned above, sexual experience can increase the volume of the preoptic area in females, whereas age can decrease the volume of the preoptic area. This points to an important principle that often is not taken into account in psychobiology or in phenotypic plasticity studies, namely that organisms age as they gain experience, but do not necessarily gain experience as they age. The only other example that we have been able to find that controlled for experience independent of age was a study of Witkin (1992) demonstrating that aging in rats is associated with a decline in the density of synaptic input to GnRH neurons in the preoptic area, but that reproductive experience will counter this trend and maintain synaptic input in old females at the levels of young adults.

Finally, the group of Crews has discovered that sexual experience can reorganize the functional associations between brain nuclei. Jon Sakata has developed methods to analyze correlations between cytochrome oxidase levels in different brain nuclei and combining these results with knowledge of neuroanatomical pathways (Sakata et al., 2000). Such analysis shows that among some nuclei sexual experience has no influence on the strength of neural connections but among other nuclei the functional associations are altered completely.

Taken together, such results indicate (i) that the volume and metabolic capacity of specific brain regions are dynamic in adulthood, changing as individuals age and gain sociosexual experience, (ii) that the size and activity of brain areas can be independent, and (iii) the embryonic environment influences the nature and degree of these changes.

## Social experience in mice

It is common in these types of studies for the sex ratio of the litter to be balanced such that there are equal numbers of male and females and to contrast them with single sex litters. But in nature sex ratios in the litter vary. This is particularly important when considering that the behavioral phenotype of knockout mice is often interpreted as the effects of the absence of the gene product on adult behavior. Could these behavioral differences among genotypes be exaggerated or blurred by the postnatal environment? That is, since mice develop in litters of varying sex ratios and genotypes, it is possible that some of these behavioral differences may result from the unique composition of the litter. To determine if these factors might play a role in the development of the behavioral characteristics that have become diagnostic of knockout mice, Crews *et al.* (2004) sexed and genotyped within two days of birth pups derived from mating of males and females heterozygous (HTZ) for a null mutation of estrogen receptor $\alpha$ (ERKO). Litters were then reconstituted, forming same-sex/mixed-genotype litters of equal numbers of ERKO and wildtype (WT) individuals. In this manner any effect of sex could be dissociated from any effect of genotype. As adults, ERKO and WT individuals were tested in a standard resident-intruder paradigm. The results indicated that the behavioral differences between the genotypes were more sharply defined than reported previously. ERKO females displayed only aggressive behavior whereas their WT littermates displayed only mounting behavior; in ERKO males both aggression and mounting behavior was greatly reduced (Fig. 4.6). These data suggest that the postnatal environment such as litter composition may influence the development of sociosexual behaviors.

## Social experience in avian species

As we have seen above mothers can adjust the competitiveness of their offspring by providing them embryonic exposure to her androgens. In addition, she may influence competitive behavior of her young by providing them a certain social environment after hatching. The effect of social experience comes about partly in interaction with hormonal factors. Social factors may stimulate hormone production, influencing the performance of social behavior, while hormones may bring the animal in a situation in which it is able to gain social experience. This interaction may influence the form, application, and

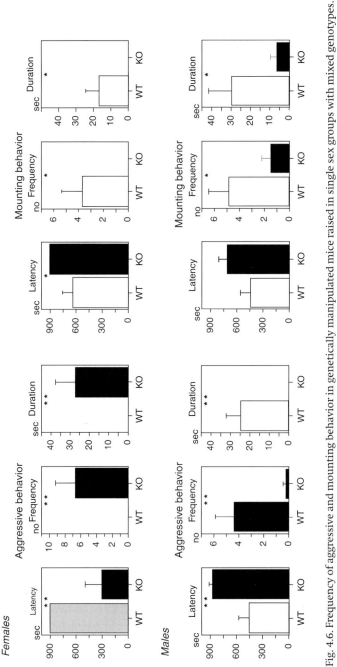

Fig. 4.6. Frequency of aggressive and mounting behavior in genetically manipulated mice raised in single sex groups with mixed genotypes. Tests with female mice involved ovariectomized female intruders (top panel) and with male mice involved olfactory-bulbectomized male intruders (bottom panel). Mice were genotyped within two days of birth and the litters reconstituted to contain equal numbers of wildtype (WT) or knockout (KO) male or female mice. Values are group mean and standard errors. Statistical analysis was computed on log-transformed data; one-tailed $t$ tests. ($^*$ $P < 0.05$, $^{**}$ $P < 0.01$. From Crews et al. (2004).)

frequency of social behavior. We will discuss these three aspects briefly in this order.

The motor coordination for the basic form of species specific postural and vocal displays in gulls and other non-oscine avian species is present already early in ontogeny, and can be activated precociously by early testosterone treatment even without substantial social experience or practice (Groothuis and Meeuwissen, 1992). Nevertheless, details of these motor patterns such as the position of the bill or wings can be shaped by social experience, probably due to operant conditioning, based on the effect a display has on its opponent (Groothuis, 1992). This exemplifies that an environmental factor that is not indispensable for normal development can still affect development when present. This shows the invalidity of "kaspar hauser" experiments, often used at the time of the nature-nurture debate at the time of Tinbergen. Such experiments, in which animals are reared in social isolation, can only tell us which factors are not indispensable for normal development, but not how normal development takes place, let alone the influence of genetic factors that are not even manipulated in such an experiment.

However, such isolation experiments have shown that social experience is indispensable for the development of the proper use of threat and courtship displays. Unfortunately, complete social isolation is such a crude manipulation that the cause of abnormal behavior is difficult to interpret. Groothuis refined this approach, and reared gull chicks in small sibling groups. Since in such groups hardly any aggressive interactions take place, Groothuis could specifically deprive the young from having such interactions without refraining them from other social input. Such birds did not show the gradual shift from overt aggression to the use of threat display that normally reared birds' show during their first year. This indicates that birds have to gain experience with social interactions to develop efficient communication behavior (Groothuis, 1992, 1993).

Finally, the frequency of social behavior is obviously depending on the number of social interactions, which in turn stimulates androgen production, facilitating social behavior. A central framework for the social regulation of androgen production is the challenge hypothesis (Wingfield et al., 1990). Assuming that testosterone imposes costs to the animal, circulating levels of this hormone should only be elevated during or in anticipation of a social challenge. The laboratory of Groothuis recently demonstrated for the first time that this is the case in young birds too (Ros et al., 2002). Black-headed gull chicks that had to defend their territory more frequently had elevated levels of the hormone, and also produced short lasting peaks of testosterone contingent on a social challenge (an artificial intruder at the territory). Interestingly, birds that experienced elevated levels of testosterone in the second week after hatching became increasingly sensitive to the hormone and this effect lasted at least until fledging. This demonstrates, in addition to the effect of embryonic exposure discussed above, another long lasting effect of early testosterone exposure. Thus, in black-headed gulls, mothers can influence the level of social stimulation her chicks receive, and thereby their early exposure to testosterone, by choosing her nest site. Chicks reared in higher densities will experience higher levels of social stimulation and thereby higher levels of androgens (Ros, 1999).

SUMMARY AND CONCLUSIONS

Based on Tinbergen's view of the study of behavioral development we have described the importance of, and some recent advances, in this field. We argue that the study of behavioral development should combine both proximate and ultimate approaches and can help to understand how early subtle environmental factors shape consistent individual variation both between and within sexes. This is illustrated by reviewing the profound effects of incubation temperature on the development of brain and social behavior in the leopard gecko, a species with temperature determined sex determination, and the effects of early exposure to steroid hormones on social behavior in rodents and especially birds. Both are maternal effects: incubation temperature can be partly determined by the nest site were the mother deposited her eggs, while in both oviparous and viviparous vertebrates maternal hormones reach and influence the embryo. In the gecko, incubation temperature affects sexual and aggressive behavior, growth, the hypothalamus-pituitary-gonadal axis, as well as the size, connectivity, and metabolic capacity of certain brain areas. In this way not only is the gonad type determined, but so too is the morphological, physiological, neural, and behavioral phenotype established that explains much of within-sex variation. In rodents, maternal hormones affect similar aspects. In avian species, maternal hormones, deposited in the eggs vary systematically between and within clutches and have both short and long lasting effects on competitive behavior. Evidence suggests that mothers adaptively adjust hormone allocation to the environmental context. In addition, we discuss some effects of postnatal experience on behavioral development in *both* gecko, mice and bird species. Our results also illustrate how the study of animal models other than rodents can help to understand important developmental processes.

We hope to have demonstrated that within the field of ethology the question about behavioral development is still a question in its own right. The field has progressed enormously since Tinbergen, and has made the nurture-nature debate clearly obsolete. The data presented in this chapter emphasize how subtle influences from the embryonic environment strongly influences species-typical behavior and are often mediated by maternal effects. Clearly, research in the area of behavioral development has left the stage of black-box analyses and now deals with the level of physiological processes and gene expression. In addition we are now much more aware that such research should include both proximate and ultimate approaches. Finally, we hope to have convinced the reader that we should now not only focus on the developmental pathways for species-typical behavior, but also try to explain variation within the same species and sex.

JERRY A. HOGAN AND JOHAN J. BOLHUIS

## 5

# The development of behavior: trends since Tinbergen (1963)

Prior to 1963, classical ethological theory had surprisingly little to say about the development of behavior, in spite of the fact that Lorenz himself had published a landmark paper on the concept of imprinting in 1935. Tinbergen's (1951) book *The Study of Instinct*, for example, has only one short chapter on development, and only one paragraph on imprinting. Lehrman (1953), in his influential critique of ethological theory, pointed out this neglect of developmental questions, which subsequently led a number of ethological workers to consider problems of development. It is likely that it also led Tinbergen (1963), ten years later, to declare ontogeny as one of the four major problems of behavioral biology.

Tinbergen's discussion of ontogeny implies it to be the study of causal mechanisms: "I should like to characterize the *phenomenon* [i.e. ontogeny] as 'change of behaviour machinery during development'. This is not, of course, the same as a change of behaviour during development;" (1963, p. 424). As Crews and Groothuis (2005, this volume) rightly point out, all of Tinbergen's "why" questions (i.e., causation, survival value, and evolution) can and should be asked of behavioral development. Nonetheless, in this chapter, we will concentrate on Tinbergen's original concern with the causal mechanisms of development, and only mention functional and evolutionary aspects briefly in the concluding section. We will also primarily consider studies using behavioral techniques.

By the mid 1960s, there was already a considerable body of research on the development of animal behavior. A group of workers at Cambridge University had studied aspects of imprinting (Hinde, 1962; Bateson, 1966) and of bird song learning (Thorpe, 1961), and Kruijt (1964) had published an extensive monograph on the development of social behavior in junglefowl. As well, a number of American

*Tinbergen's Legacy: Function and Mechanism in Behavioral Biology*, Johan J. Bolhuis and Simon Verhulst (eds.). Published by Cambridge University Press. © Cambridge University Press 2008.

psychologists had addressed developmental problems in their research (for a review see Lehrman, 1970). However, some of the significance of this work was overshadowed by the nature/nurture debate at that time. In this chapter, we begin with a consideration of that debate, and then very briefly review more recent work on filial and sexual imprinting, bird song learning, behavior systems in general, human language development, and attachment in humans. We conclude with some observations on trends in the field.

## THE NATURE/NURTURE DEBATE

In his early papers, Lorenz (1937) postulated that behavior could be considered a mixture of innate and acquired elements (*Instinkt-Dressur-Verschrankung*: intercalation of fixed action patterns and learning), and that analysis of the development of the innate elements (fixed action patterns) was a matter for embryologists. In reaction to Lehrman's (1953) critique of ethological theory, Lorenz (1965) changed his formulation of the problem, and argued that the information necessary for a behavior element to be adapted to its species' environment can only come from two sources: from information stored in the genes or from an interaction between the individual and its environment. This formulation also met with considerable criticism from many who insisted that development consisted of a more complex dynamic. Gottlieb (1997) discusses many aspects of this debate, and we have recently republished some of the original papers (Bolhuis and Hogan, 1999). Crews and Groothuis (2005, this volume) address Tinbergen's discussion of ontogeny in some detail. Here, we will mention only two important issues.

First, Lehrman (1970) pointed out that he and Lorenz were really interested in two different problems: Lehrman was interested in studying the effects of all types of experience on all types of behavior at all stages of development, very much from a causal perspective, whereas Lorenz was interested only in studying the effects of functional experience on behavior mechanisms at the stage of development at which they begin to function as modes of adaptation to the environment. Thus, Lehrman used a causal criterion to determine what was interesting to study, while Lorenz used a functional criterion. Hogan (1988) notes that both these viewpoints are equally legitimate, but that Lorenz's functional criterion corresponds to the way most people think about development. We return to this point in our concluding section.

The second issue is that even behavior patterns that owe their adaptedness to genetic information require interaction with the environment in order to develop in the individual. As Lehrman states: "The interaction out of which the organism develops is *not* one, as is so often said, between heredity and environment. It is between *organism* and environment! And the organism is different at each state of its development." (Lehrman, 1953, p. 345). This view has been developed more recently by Oyama (1985), and has gradually been adopted by most students of behavior. Several of our examples in later sections clearly illustrate this interactionist interpretation of development.

### FILIAL IMPRINTING: SENSITIVE PERIODS, IRREVERSIBILITY, AND PREDISPOSITIONS

Filial imprinting is the process through which early social preferences become restricted to a particular class of stimuli (usually the mother and siblings) as a result of exposure to those stimuli. Lorenz (1935) originally proposed that such exposure would be effective only during a brief critical period shortly after birth (or hatching) and that the preferences formed could not be reversed once imprinting had occurred. Subsequent research showed these generalizations to be oversimplified (for reviews see Bateson, 1966 and Bolhuis, 1991). We will mention here several issues raised in studies on filial imprinting that also have implications for developmental studies in general.

The concept of a critical period, which Lorenz borrowed from embryology, implies that the animal can be greatly influenced by certain types of experience at a particular phase of its development, but is less sensitive to the same experience both before and after that phase. Many studies have shown that the phase of greatest sensitivity in many species is not as restricted as Lorenz originally supposed, and Bateson and Hinde (1987) have argued that the term sensitive period is much better to describe such phases of development; this term is now widely, but by no means universally, used.

What are the causes of sensitive periods? The conventional view is that the causes are some sort of physiological clock mechanism (see reviews in Bornstein, 1987; Rauschecker and Marler, 1987). In filial imprinting, it is likely that the onset of the sensitive period can be explained in terms of developing physiological factors such as increases in visual efficiency and motor ability in precocial birds some time after hatching. However, different causal factors are thought to be involved at the end of the sensitive period. Sluckin and Salzen (1961) suggested that the ability to imprint comes to an end after the animal has developed a social preference for a certain stimulus as a result of exposure to that stimulus. The animal will stay close to the familiar object and avoid novel ones; it will thus receive very little exposure to a novel stimulus and there will be little opportunity for further imprinting. This interpretation implies that imprinting will remain possible if an appropriate stimulus is not presented. Indeed, chicks that are reared in isolation retain the ability to imprint for longer than socially reared chicks. An apparent decline in sensitivity with age in isolated chicks can be explained as resulting from the animals' imprinting to stationary visual aspects of the rearing environment (Bateson, 1964). Thus, it is the imprinting process itself that brings the sensitive period to an end.

Bateson (1987) has proposed a competitive exclusion model that suggests how such a process might work. Neural growth is often associated with particular sensory input from the environment. Bateson's model assumes that there is a limited capacity for this neural growth to impinge upon the systems that are responsible for the execution of the behavior involved (e.g., approach, in the case of filial imprinting). Input from different stimuli competes for access to these executive systems. Once neural growth associated with a certain stimulus develops control of the executive systems, subsequent stimuli will be less able to gain access to these systems. More recently, Bateson and Horn (1994) have formalized this model using neural networks. Further, insofar as these early neural connections are permanent (see Shatz, 1992; Hogan, 2001, pp. 263–269), this interpretation also explains the general irreversibility of many aspects of early learning. We describe later evidence from studies of sexual imprinting and bird song learning that is also consistent with an experience-dependent end to the sensitive period. These results all show that it is necessary to investigate the causes for both the beginning and the end of any sensitive period before reaching conclusions about the mechanisms responsible.

Van Kampen (1996) has proposed a framework for the study of filial imprinting that is based upon the behavior system concept of Hogan (1988, 2005, this volume). This scheme, shown in Fig. 5.1, makes clear that a necessary component in understanding filial imprinting is studying the development of the perceptual mechanism that is responsible for recognition of the imprinting stimulus (S in the figure). The process of combining features to develop a stimulus-recognition mechanism can also be called the formation of an internal representation of the imprinting stimulus. Van Kampen reviews a large number of studies by him and his colleagues showing that the development of the perceptual mechanism for recognizing the imprinting stimulus

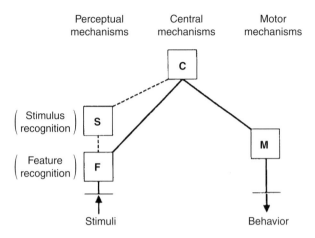

Fig. 5.1. Basic scheme of the filial system of a chick. Perceptual mechanisms include feature recognition in F (such as color, shape, size, and movement) and stimulus recognition in S (representations of stimuli on which a chick is imprinted). The motor mechanisms in M are responsible for the execution of motor patterns such as shrill calling, searching, approach, and twitter calling. The central mechanisms in C can be observed as the internal state of the chick, for instance its motivation to seek contact with the imprinting stimulus. (From van Kampen, 1996. Copyright 1996 by the Psychonomic Society.)

follows the same rules of associative learning as found in studies of perceptual learning in general (see also Honey and Bolhuis, 1997). It can be noted that Bateson's model of competitive exclusion discussed above does not include the central mechanism depicted in Fig. 5.1. Rather, in his model, the developing representation of the imprinting object connects directly to the motor mechanisms. Van Kampen's (1996) review, however, provides considerable evidence that the filial system of a chick must contain at least two central mechanisms, which he calls a search mechanism and a contentment mechanism.

Other studies of the development of the filial system in young, precocial birds have shown that there must also be at least two perceptual mechanisms that are neurally and behaviorally dissociable. One recognizes the imprinting object and the other responds to general features of conspecifics; the latter mechanism has been called a predisposition (Horn, 1985, 1998; Bolhuis, 1996; Bolhuis and Honey, 1998). The development of the predisposition is especially interesting because training with a specific stimulus is not necessary for it to emerge: the predisposition can emerge in dark-reared chicks, provided that they receive a certain amount of non-specific stimulation, such as an opportunity to run in a wheel or being handled, within a certain period during development (Johnson et al., 1989). Thus, the predisposition develops prefunctionally because the chick approaches specific visual stimuli even though it has not had any visual experience; nonetheless, this effect is not seen if the chick does not have other kinds of experience. A similar example is provided by the development of the auditory recognition mechanism of the species' maternal call in Peking ducklings (see below). The stimulus characteristics of the visual stimuli that allow the filial predisposition to be expressed were investigated in tests involving an intact stuffed junglefowl versus a series of increasingly degraded versions of a stuffed junglefowl. The intact model was preferred only when the degraded object possessed no distinguishable junglefowl features (Johnson and Horn, 1988). Further studies showed that the necessary stimuli are not species- or even class-specific: Eye-like stimuli are normally important, but other aspects of the stimulus are also sufficient for the expression of the predisposition (Bolhuis, 1996).

There are interesting similarities between the development of face recognition in human infants, and the development of filial preferences in chicks (Blass, 1999; Johnson and Bolhuis, 2000). Newborn infants have been shown to track a moving face-like stimulus more than a stimulus that lacks these features, or in which these features have been jumbled up. Similarly, in both human infants and young precocial birds, the features of individual objects need to be learned. Once learned, both infants and birds react to unfamiliar objects with species-specific behavior patterns (van Kampen's search mechanism) that tend to bring them back to the familiar object or caregiver (van Kampen's contentment mechanism).

One final aspect of the developing filial system deserves mention. In an extensive series of experiments published between 1975 and 1987, Gottlieb (for review see 1997) investigated the mechanisms underlying the preferences that young ducklings of a number of species show for the maternal call of their own species over that of other species. He found that differential behavior toward the species-specific call could already be observed at an early embryonic stage, before the animal itself started to vocalize. However, a post-hatching preference for the conspecific maternal call was only found when the animals received exposure to embryonic contact-contentment calls, played back at the right speed and with a natural variation, within a certain period in development. Thus, the development of the perceptual mechanism responsible for recognizing the maternal call in ducklings is dependent on specific kinds of experience before hatching.

Lorenz (1935) thought that imprinting had "nothing to do with learning." Indeed, filial imprinting proceeds without any obvious reinforcement such as food or warmth (see Bolhuis *et al.*, 1990, for discussion). However, an imprinting object may itself be a reinforcer, that is, a stimulus which an animal finds rewarding. Just as a rat can learn to press a lever to receive a food reward, so a visually naive chick is able to learn to press a pedal to see an imprinting object. This is a case of instrumental or operant conditioning, and in the case of the chick the mere sight of an imprinting object is reinforcing the response. Similarly, when chicks are exposed to two imprinting stimuli (e.g., a visual object and a sound) simultaneously, they learn more about the individual stimuli than when they are exposed to the stimuli sequentially, or to only one stimulus (Bolhuis and Honey, 1994). This so-called within-event learning has also been found in conditioning paradigms in rats and humans (Bolhuis and Honey, 1998). So, although imprinting looks quite different from classical or operant conditioning, nevertheless it shares many characteristics with these forms of associative learning. Thus it may be that only the characteristics and the circumstances in which imprinting occurs differ from other forms of learning, but that the underlying mechanisms are similar, if not the same, a suggestion that has also been put forward in the context of sexual imprinting, as we shall see in the following section.

### SEXUAL IMPRINTING: REPRESENTATIONS AND STABILITY

Sexual imprinting is the process by which early experience affects sexual preference later in life. Although sexual imprinting was one of the earliest subjects studied by ethologists (Lorenz, 1935), there have been many fewer studies of it than of filial imprinting, primarily because experiments require very much more time to do, and proper control conditions are difficult to arrange. Early studies were relatively limited and anecdotal (for review see Immelmann, 1972), but more recently the investigation of sexual imprinting has been conducted in a systematic fashion. In sexual imprinting experiments, birds are usually reared up to a certain age with parents of their own species, or with foster parents of a different species. When the animals are sexually mature, their sexual preferences are tested. Measures of preference include time near the test stimuli, amount of courtship shown, and copulation choices.

In one early study, Schutz (1965) raised mallard ducks (*Anas platyrhynchos*) with a variety of foster species. He reported that, although most mallard males tended to mate with a female of the foster species with which they were reared, a minority of males still preferred to mate with conspecific females. Moreover, female mallards preferred to mate with a conspecific male, even when reared with another species. Schutz interpreted these data as evidence for a species-specific response bias. However, Kruijt (1985) pointed out that the females in Schutz' study were released onto a large pond where they were likely to have been exposed to directed courtship of males that had been reared with their own species. This experience could have been responsible for the females' choice. In experiments to test this hypothesis, Kruijt *et al.* (1982) raised female mallards with individuals of either a white or a wild strain. The females were later tested in a simultaneous choice situation between a white and a wild male. Unlike the females in Schutz' study, most females chose the male of the strain with which they had been reared. In those cases in which the female chose a male of the opposite strain, it was determined that the chosen male directed more courtship toward the female than the male of the same strain. Kruijt concluded that early experience did determine the females' preference, and that females respond to activity differences of males. Thus, there was no evidence for a conspecific bias in this case.

A conspecific bias in sexual imprinting was also suggested for zebra finch (*Taeniopygia guttata*) males (Immelmann, 1972). The work of ten Cate and his collaborators, however, has demonstrated that the formation of sexual preferences in zebra finch males is more complex (ten Cate, 1984; for reviews see 1989, 1994). In a series of studies, zebra finch young were cross fostered with a mixed pair of parents, a zebra finch and a Bengalese finch (*Lonchura striata*). In preference tests in which zebra finch males were given a simultaneous choice between a female zebra finch and a female Bengalese finch, ten Cate replicated the findings of Immelmann: zebra finch males preferred to court females of their own species when adults. However, analysis of the behavior of the parents during the first 55 days of the young birds' life revealed that zebra finch foster parents spent significantly more time with the young in a number of social behaviors compared to the Bengalese finch foster parents. Later studies in which the amount of social interaction between parent and young was manipulated showed that the less the interaction with the zebra finch parent, the greater the sexual preference for the Bengalese female. In other studies (Kruijt *et al.*, 1983), zebra finch males were raised with two Bengalese parents.

In this case almost all the zebra finch males preferred a Bengalese female. However, the presence of zebra finch siblings during rearing significantly shifted the preference of the males toward the zebra finch female. Thus, the eventual sexual preference of zebra finch males was affected by the social interactions that the young bird experienced with parents and siblings during development. Once again, it was not necessary to invoke species-specific predispositions to explain the behavior.

As with filial imprinting, a necessary component for understanding sexual imprinting is studying the development of the perceptual mechanism responsible for recognition of the imprinting stimulus. Ten Cate (1987) examined the sexual preferences of zebra finch males raised by zebra finch parents for the first 30 days, then placed individually in groups of Bengalese finches for the next 30 days, and thereafter raised in social isolation. Under these conditions many of the birds will court females of both species when tested after sexual maturity; such birds are considered to be double imprinted. Ten Cate was interested to discover whether the double-imprinted birds had developed two separate internal representations (perceptual mechanisms) of the rearing species, or had developed one internal representation that combined the imprinted information of the two rearing species. He gave the birds preference tests in which zebra finch, Bengalese finch, and hybrids of the two species were used. The results showed the female hybrids to be more attractive than females of either species. This outcome suggests that zebra finch males combine the imprinted information of the two rearing species into one internal representation. A review of the relevant literature (Hollis et al., 1991) led to the conclusion that the rules by which combinations of stimulus features are represented in memory are the same for the formation of the stimulus-recognition mechanisms studied in imprinting studies as for associations formed in perceptual learning.

The final aspect of sexual imprinting that we will consider is the stability of sexual preferences. It was long thought that once a significant sexual preference was established, it would remain stable (Lorenz, 1935; Immelmann, 1979). Under some conditions, this is clearly not the case. In two independent studies (Immelmann et al., 1991; Kruijt and Meeuwissen, 1991, 1993), male zebra finch young were reared with Bengalese finch foster parents for 40 days and subsequently isolated from them until day 100 (sexual maturity). Half of the males were then given a preference test and all showed a strong preference for females of the foster species. All the birds then received breeding experience with a female of their own species for several months. Thereafter, the birds were tested in two series of preference tests, one immediately after the end of the breeding experience, the other some months later. The results of both studies were remarkably similar. Whereas in the group that received a pretest, most of the birds retained their original preference for the foster species, the majority of the males in the group without pretest preferred females of their own species. Results such as these led Bischof (1979, 1994) to suggest that there was a second sensitive period at sexual maturity that led to either consolidation or modification of the perceptual mechanism for recognition of the appropriate sexual partner depending on the animal's first sexual experience. Bischof thus describes sexual imprinting as a two-stage process.

### BIRD SONG LEARNING

Many years ago, Thorpe (1961) showed that the male chaffinch, *Fringilla coelebs*, had to learn to sing its species-specific song, and that this learning occurred in two stages. First the young bird had to hear the normal song (or, within limits, a similar song); later it learned to adjust its vocal output to match the song it had heard when it was young. Similar results have also been found for the white-crowned sparrow, *Zonotrichia leucophrys* (Konishi, 1965; Marler, 1970a), though not necessarily for other species of song birds (Marler, 1976; Logan, 1983; Hultsch, 1993).

The first, or memorization, stage of learning involves the development of a perceptual mechanism. Konishi (1965) and Marler (1976, 1984) proposed that the results of studies of song learning imply the existence of an auditory template, which was conceived of as a sensory mechanism that embodies species-specific information. The normal development of the template requires auditory experience of the proper sort at the proper time. In our terms, the template becomes a song-recognition (perceptual) mechanism that is partially formed at hatching. One question that has been asked is whether there is one or many templates (cf. discussion of internal representations in sexual imprinting above). Originally it was thought that the young bird memorized a single song and that later variation in produced song came about because of mismatches during the selection stage. More recently (Marler and Peters, 1982; Nelson, 1997), it has become clear that the bird memorizes a variety of species-specific songs when young, but only one (or one subset) of these is selected later for production. How this choice is made is not known, but it now appears that the neural substrate for the template used in the development of song production is located in a different part of the brain from the neural substrate subserving the other song memories (Jarvis and Nottebohm, 1997; Bolhuis and Eda-Fujiwara, 2003; Bolhuis and Gahr, 2006; Gobes and Bolhuis, 2007; Bolhuis, 2008, this volume). The concept of templates raises the question of modularity in the central nervous system, and this question will be considered again in the concluding section.

A second question concerns constraints on the kinds of experience that can affect development. Thorpe (1961) found that chaffinches would learn to sing normal or rearranged chaffinch songs heard when young, but exposure to songs of other species resulted in songs no different from those sung by birds raised in auditory isolation. When fledgling male song sparrows (*Melospiza melodia*) and swamp sparrows (*Melospiza georgiana*) were exposed to taped songs that consisted of equal numbers of songs of both species, they preferentially learned the songs of their own species. Males of both species are able to sing the songs of the other species; thus it appears that they are predisposed to perceive songs of their own species; Marler (1991) called this "the sensitization of young sparrows to conspecific song" (p. 200). The range of stimuli that affect development turns out to depend crucially on such factors as the species, the age at which the bird is exposed, the previous experience of the bird, and the conditions under which the bird is exposed (Marler, 1987; Slater *et al.*, 1988; Catchpole and Slater, 1995; Nelson, 1997). There are no easy causal generalizations.

A third question concerns the processes that are involved in development. In the memorization stage, it is often assumed that mere exposure to an adequate stimulus is sufficient for perceptual learning to occur. In a restricted sense, this is probably true, but what makes a stimulus adequate often depends critically on the conditions under which the bird is exposed: For example, in many cases, memorization is more likely to occur when exposure occurs during social interaction with another bird (Baptista and Petrinovich, 1984; Petrinovich, 1985; Clayton, 1994; Baptista and Gaunt, 1997; Nelson, 1997), though the mechanism through which social interaction has these effects remains an open question (Houx and ten Cate, 1998, 1999).

The development of the perceptual mechanism for song recognition generally occurs many weeks or months before the bird actually learns to sing its song. Learning to sing occurs when the bird's internal state (e.g., the level of testosterone) is appropriate. At this point, the bird learns to adjust its motor output to match the image it has formed previously. This adjustment must involve the bird's hearing itself because deafened birds never learn to produce any song that approaches normal song (Konishi, 1965). Experiments by Stevenson (1967) showed that hearing its species-specific song could serve as a reinforcer for an operant perching response in male chaffinches. On the basis of these results, Hinde (1970) suggested that song learning might involve matching the sounds produced by the young bird with the stored image: sounds that matched the image would be reinforced, whereas other sounds would be extinguished. In this way, a normal song could develop in much the same way as an experimenter originally trains a rat to press a lever (Skinner, 1953).

In most species, three stages in the production of song can be distinguished: subsong, plastic song, and crystallized song (Thorpe, 1961). During the subsong phase, the bird essentially produces what in human infants is called "babbling," and slowly adjusts its production to match phrases and songs it heard during the memorization stage; it may also invent new combinations of phrases during this phase. These changes presumably come about in the manner suggested by Hinde. In the plastic song phase, the bird may be singing a number of songs that resemble songs it previously heard. Which of these songs becomes chosen as the crystallized song depends, in many species, on the songs it hears from other birds at this time. In some species, the bird selects a similar song (which probably accounts for the occurrence of local dialects) and in other species, the bird selects a dissimilar song. In either case, a selection is made from songs already developed (Marler and Peters, 1982; Nelson, 1997). The selection process presumably also involves some kind of reinforcement, often provided by the behavior of conspecifics. A particularly interesting example is the song of the brown-headed cowbird, *Molothrus ater*. Males of this species increase their performance of those songs that are associated with a wing stroke display given by the females (West and King, 1988, 2001). Once the song has crystallized, it has long been thought that auditory feedback from singing was no longer necessary to maintain the song, at least in most species. It now appears that auditory feedback after crystallization continues to affect the song in most species, and is necessary for the song in many species (Brainard and Doupe, 2000).

Although most research on song learning has focused on males, there is now considerable evidence that females also develop song-recognition mechanisms (for review see Riebel, 2003b; Bolhuis and Eda-Fujiwara, 2003). For example, Riebel (2000) found that early exposure of female zebra finches to the song of a zebra finch male led to a preference for that song over the song of an unfamiliar male when the females were tested in adulthood. Riebel *et al.* (2002) found that such female song preferences are comparable to song preferences in males. As with male zebra finches, there is a sensitive period between 35 and 65 days of age for this exposure to be effective (for review see Riebel, 2003a).

## DEVELOPMENT OF BEHAVIOR SYSTEMS

Kruijt (1964), in his classic monograph on the development of social behavior in the junglefowl (*Gallus g. spadiceus*), suggested that in young junglefowl chicks – and obviously, in the young of other species as well – many of the motor components of behavior appear as independent units prefunctionally, that is, prior to any opportunity for practice. Only later, often after specific experience, do these motor components become integrated into more complex systems such as hunger, aggression, and sex. Hogan (1988) has generalized this proposal by Kruijt and suggested a framework for the analysis of behavioral development using the concept of behavior system, which comprises perceptual, central, and motor mechanisms that act as a unit in some situations. This concept has been described more fully above (see Hogan, 2005, this volume). Behavioral development is essentially the development of these mechanisms themselves and of the changes in the connections among them. Often, these mechanisms and their connections only develop after functional experience, i.e., experience with the particular stimuli involved, or with the consequences of performing specific motor patterns.

An example of a developing behavior system is the hunger system in the junglefowl chick (Hogan, 1971, 1988). This system includes perceptual mechanisms for the recognition of features (e.g., color, shape, size), objects (e.g., grains, mealworms), and functions (food versus nonfood); see Fig. 5.2. There are also motor mechanisms underlying behavior patterns such as ground scratching and pecking, and there is a central hunger mechanism. Importantly, several of the connections between the mechanisms (shown by dashed lines in Fig. 5.2) only develop as a result of specific functional experience. For instance, only after a substantial meal will the chick differentiate between food items and non-food items to eat (Hogan-Warburg and Hogan, 1981). On the motor side of the system, a young chick's pecking behavior is not dependent on the level of food deprivation before 3 days of age. Only after the experience of pecking and swallowing some solid object do the two mechanisms become connected, and only then is the level of pecking dependent on the level of food deprivation. A similar phenomenon

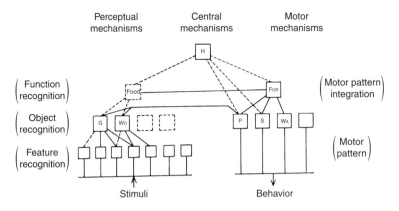

Fig. 5.2. The hunger system of a young chick. Boxes represent putative cognitive (neural) mechanisms. Perceptual mechanisms include various feature recognition mechanisms (such as of color, shape, size, and movement), object recognition mechanisms (such as of grain-like objects [G], worm-like objects [Wo], and possibly others) and a function recognition mechanism [Food]. Motor mechanisms include those underlying specific behavior patterns (such as pecking [P], ground scratching [S], walking [Wa], and possibly others), and an integrative motor mechanism that could be called foraging [For]. There is also a central hunger mechanism [H]. Solid lines indicate mechanisms and connections among them that develop prefunctionally. Dashed lines indicate mechanisms and connections that develop as a result of specific functional experience. (From Hogan, 1988. Copyright 1988 by Plenum Press.)

occurs with respect to suckling in rat pups, kittens, puppies, and human infants. In the case of rat pups, suckling does not become deprivation-dependent until about two weeks after birth (Hall and Williams, 1983). Unlike the chicks, we do not yet know what experience is needed to connect the sucking motor mechanism with the central hunger mechanism in the rat pup.

The development of behavioral structure is not uniform, but may proceed along different pathways for different behavior systems. A diagram of the dustbathing behavior system in junglefowl is presented in Fig. 5.3. (For a description of dustbathing behavior and an analysis of its causation, see Hogan, 2005, this volume.) Unlike the development of feeding behavior in rats and chicks, dustbathing is deprivation-dependent as soon as it appears in the animal's behavioral repertoire (Hogan et al., 1991). Thus, in this case, chicks do not require functional experience to connect the motor mechanisms with the central dustbathing mechanism.

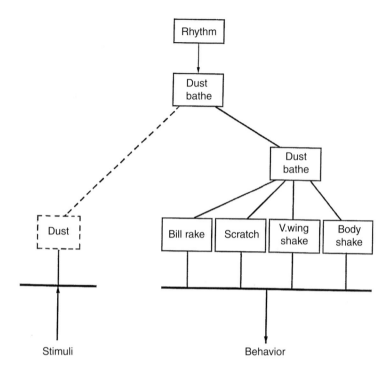

Fig. 5.3. The dustbathing system of a young chick. Boxes represent putative cognitive (neural) mechanisms: a perceptual mechanism responsible for recognizing dust, a central dustbathing mechanism responsible for integrating input from the perceptual mechanism and other internal influences as well as for coordinating output to the motor mechanisms, and several motor mechanisms responsible for the various motor patterns comprising dustbathing. The motor mechanism labeled "dustbathe" is responsible for organizing the temporal patterning of the specific motor mechanisms. Solid lines indicate mechanisms and connections that develop prefunctionally. Dashed lines indicate mechanisms and connections that develop as the result of specific functional experience. (From Vestergaard et al., 1990. Copyright 1990 by E. J. Brill.)

Larsen et al. (2000) studied in detail the development of dustbathing behavior sequences in chicks from hatching to 3 weeks of age. They found that the individual behavior elements, as soon as they appeared, were incorporated into the normal adult sequence structure; this occurred even though the form of the elements themselves is not yet fixed. These results support the conclusion that separate mechanisms are responsible for the form of the individual behavior elements and for the organization of these elements into recognizable sequences as shown in Fig. 5.3. A similar conclusion was reached by Berridge (1994) on the basis of results on the development of grooming sequences in young mice (Fentress, 1972; Fentress and Stilwell, 1973) and young rats and guinea pigs (Colonnese et al., 1996). In fact, Berridge et al. (1987) called certain sequences of grooming movement "syntactic chains" to emphasize the rules controlling natural action sequences (see also Fentress and Gadbois, 2001). We return to these ideas in the section on human language.

Functional experience plays an essential role in the development of the perceptual mechanism for recognizing dust and of the connection between it and the central mechanism (dashed lines in Fig. 5.3). Kruijt (1964) found that making the external situation as favorable as possible for dustbathing was insufficient for releasing the behavior. This result implies that early dustbathing may be controlled exclusively by internal factors (see Hogan, this volume). It also implies that the connection between the dust recognition perceptual mechanism and the central mechanism is not formed until well after the motor and central mechanisms are functional. Vestergaard and Hogan (1992) found that early dustbathing is most likely to occur in whatever substrate is pecked at most. They point out that pecking is a movement that functions as exploratory, feeding, dustbathing, and later aggressive behavior. They suggest that perceptual mechanisms specific to each system develop gradually out of exploratory pecking on the basis of functional experience (see also Chow and Hogan, 2005). It remains to be determined whether removal of lipids, the sensory feedback from the substrate in the feathers, or facilitation of the dustbathing behavior itself is the crucial factor. Other evidence shows that early experience can lead to stable preferences for particular stimuli (Petherick et al., 1995; Vestergaard and Baranyiova, 1996). As an extreme example, Vestergaard and Hogan (1992) raised birds on wire mesh but gave them regular experience on a substrate covered with coal dust, white sand, or a skin of junglefowl feathers. In choice tests given at 1 month of age, some of the birds that had had experience with junglefowl feathers were found to have developed a stable preference for dustbathing on the feathers. This example is important because it shows how a system can develop abnormally.

HUMAN LANGUAGE

Hogan (2001) has discussed how it is possible to consider human language to be a behavior system that is similar in many respects to the behavior systems we have considered already. Human language, of course, is vastly more complex than dustbathing or feeding in chickens, but, as a biological system, both the organization and development of language should share many of the principles governing these simpler systems. To begin, it is necessary to identify the building blocks of the language system: What are the perceptual, motor, and central mechanisms comprising the system? We can start with the perceptual and motor mechanisms that recognize and produce the sounds in a language (i.e., phonemes).

It has been known for some time that human infants as young as 1 month are able to perceive phonetic distinctions categorically in a similar way to normal adults (Eimas *et al.*, 1971, 1987). More recent evidence demonstrates that these perceptual categories can be altered by linguistic experience. For example, in a cross-cultural study of 6-month-old American and Swedish infants, Kuhl *et al.* (1992) found that the two groups exhibited a language-specific pattern of phonetic perception to native- and foreign-language vowel sounds. Of particular interest is that these effects of experience are seen by 6 months of age, that is, before the infant itself begins producing speech sounds. Further, by 1 year of age, infants no longer respond to speech contrasts that are not used in their native language, even those that they did discriminate at earlier ages (see Werker and Tees, 1992). Thus, the perceptual mechanisms responsible for speech perception in infants are both highly structured at birth and highly malleable in that they are shaped, instructively and selectively, by exposure to the linguistic environment (Kuhl, 1994).

Normal infants begin to babble between 6 and 10 months (for review see Locke, 1993). The initial sounds produced by the infant are species specific (i.e., are similar in infants raised in different linguistic environments), and include phonemes not found in its native language. As the child grows older, the distribution of sounds comes more closely to approximate the distribution in its linguistic environment, and the non-native sounds drop out. The mechanism by which these changes occur has not really been analyzed adequately, but it presumably involves a process of matching vocal output to the previously developed perceptual mechanisms (templates) by auditory feedback (Marler, 1976).

These results for the development of the perceptual and motor mechanisms that recognize and produce speech sounds involve the same problems of modularity, constraints, and processes that we have seen before, especially with respect to the changes that occur in the development of bird song. These parallels have been noted for many years now (e.g., Lenneberg, 1967; Marler, 1970b), and continue to provide mutual insights into the development of both systems at both a behavioral and neural level (Hauser, 1996; Snowdon and Hausberger, 1997; Doupe and Kuhl, 1999; Brainard and Doupe, 2000, 2002; Hauser *et al.*, 2002).

Speech sounds, however, are only one aspect of normal spoken language. Sounds become combined into words (morphemes: units of meaning), and one can ask whether the phonemes or the morphemes are the basic units of the language system. Words can always be broken down into their constituent sounds, but there is now considerable evidence that infants learn utterances (words or short phrases) as a whole during the first two years with respect to both perception and production (Jusczyk, 1997; Locke and Snow, 1997). It is only later that children are able to break utterances down into smaller sound packets. For these, and other, reasons, Locke (1994) argues specifically that phonemes are not the basic building blocks of human language. We might then conclude that morphemes are the basic unit of the language system, but first we must consider what morphemes or words represent.

Birds do not sing randomly. They sing when the appropriate internal and external factors are active. In most cases this is when the sexual and/or aggression behavior systems are activated. Humans speak in comparable circumstances, but the range of circumstances in which humans speak is very much broader. In fact, cognitive psychologists (Shelton and Caramazza, 1999) have proposed that humans possess a semantic system that receives input from spoken and written words (phonological and orthographic input lexicons) and responds with output of spoken and written words (phonological and orthographic output lexicons). Our speech (or writing) thus expresses the state of our semantic system. Of course, our semantic system is much more complex than the sexual and aggression systems of songbirds (and much of the field of cognitive psychology is devoted to understanding the organization of the semantic system and the mechanisms of lexical access to it), but it seems certain that the principles of organization and development are similar. In the present context, some of Shelton and Caramazza's conclusions are particularly interesting. They reviewed studies of language processing following brain damage and found results that "broadly support a componential organization of lexical knowledge – the semantic component is independent of phonological and orthographic form knowledge, and the latter are independent of each other." (1999, p. 5) In our terms, their language system can be considered to have a central semantic mechanism with perceptual mechanisms for recognizing words and motor mechanisms for producing words.

One important component of language still has not been discussed: Words can be combined into sentences, and it is at this level of analysis that the concepts of grammar and syntax are generally used. It is also this level of organization to which the concept "language instinct" (Pinker, 1994) has been applied. The operation of grammatical structures is not normally apparent until some time after the age of 2 years, when words become recombined into novel utterances that follow particular rules (Locke and Snow, 1997). One set of rules, called Generative Grammar, was proposed by Chomsky (1965). These rules have been reasonably successful in describing the types of sentences produced by native speakers of English, but there has been great controversy about how these rules develop in the child, particularly whether specific kinds of linguistic experience are necessary (see, e.g., Tomasello, 1995). The details of this controversy need not concern us here except to say that most of the issues are the same as we have already met in describing the development of grooming (Berridge, 1994) and dustbathing (Larsen *et al.*, 2000) sequences, some of which are considered further in the concluding section.

As a final point, there is a long history of authors' proposing uniqueness for the human species on the basis of aspects of the syntax found in our language. In a recent discussion of this issue, Kako (1999) continues this debate. He points out that in its "most generic form, syntax is defined as a set of rules for assembling units of any type into larger units." He then discards this definition because he finds it too broad, and proposes instead a set of four structural properties that define the core of syntax. Once again, the details of his proposal need not concern us here. It is our opinion, however, that the generic definition has many advantages if one is interested in looking for the similarities, rather than the differences, among systems (cf. Lashley, 1951). It must be true, by definition, that human language is unique, because all species are unique, as are the specific behavior systems possessed by each species.

These results suggest that the human language system comprises three basic sets of components at two major levels of organization, and that these components develop largely independently. The sensory-motor components correspond to the perceptual and motor mechanisms (with additional connections between them and the central mechanisms), whereas the semantic (meaning) and syntax components correspond to two separate central mechanisms.

This general conception is also supported by the results of studies of deaf children. For example, deaf children born to deaf parents who communicate using sign language do not babble vocally; rather, such children babble with their hands (Petitto and Marentette, 1991). Manual babbling occurs at about the same age that vocal babbling occurs in children with normal hearing who have been raised in a vocal environment. Further, the development of sign language proceeds in much the same way as the development of vocal language, with respect to both structure and use. Goldin-Meadow (1997) has found that the same general rules apply and that they appear at the same age. These results all suggest that the language system can use auditory–vocal units or visual-manual units equally well. Studies of the neural organization of language (Hickok *et al.*, 1998) are also consistent with this interpretation.

One can ask, finally, whether this conception of human language as a behavior system actually furthers our understanding of language and its development. We think it does in at least two important ways. First, by breaking the system up into its components, the study of the pieces becomes more tractable. There has already been considerable success in comparing the development of bird song and human speech (Doupe and Kuhl, 1999), and the development of grooming sequences may provide a useful model for some aspects of the development of syntax. Further, insofar as these components are the "natural" pieces of the system, it becomes easier to understand how the system could have evolved (cf. Pinker, 1994; Hauser, 1996; Hauser *et al.*, 2002). A second important reason is that development of all three sets of components requires both functional and non-functional experience, and involves the same problems of modularity, constraints, and processes that have appeared before: Solutions to these problems in one system should easily generalize to other systems.

### ATTACHMENT BEHAVIOR

Ethological theories and methods have played an important role in the formulation and development of John Bowlby's (1969, 1991) theory of attachment in humans. This theory was originally developed to explain the behavior of children that had been separated from their mother and raised in a wartime nursery during the Second World War, and was greatly influenced by Lorenz's ideas about imprinting. This is not the place to discuss attachment theory in detail, but we can point out that in many ways, the attachment system postulated by Bowlby is analogous to the filial behavior system in young birds. In both cases, the newborn infant or chick possesses a number of behavior patterns that keep it in contact with the parent (or other caregiver) and that attract the attention of the parent in the parent's absence. Further, both infant and chick must learn the characteristics of the parent, which is considered to be the formation of a bond between the two. Factors influencing the formation of the bond are also similar, including all the factors we have discussed above, such as length of exposure, sensitive periods, irreversibility, and predispositions. Studying the importance of these factors in the human situation has resulted in a large body of literature, some of which has supported the theory and some not (Rutter, 1991, 2002). The theory itself has been modified to take these results into account, and has also been expanded to include development of attachments throughout life.

Gathering data to test hypotheses about human behavior always presents special challenges because of the ethical issues involved. To study the effects of maternal separation on infant behavior, Harlow (1958), for example, raised infant rhesus monkeys in complete social isolation, which led to horrific effects on the infant's subsequent behavior. Less intrusive methods such as raising infants with other infants (Harlow and Harlow, 1962), or separating infants from their mothers for brief periods of time (Hinde and Spencer-Booth, 1971; Hinde, 1977) led to less dramatic changes in behavior, but these methods are still unacceptable for human research. Bowlby felt that the best method for studying human development was to observe infants in real-life situations, in much the same way as many ethologists study the behavior of other animals in natural or semi-natural settings. Much of his theorizing about human attachment was based upon such research carried out by Mary Ainsworth. She and her colleagues (1978) developed a standardized "strange situation" test in which a stranger approaches an infant with and without the parent being present, and various aspects of the infant's behavior are measured. This method is now widely used, and has allowed researchers to characterize specific patterns of attachment and their determinants. Use of basically similar methods has allowed results from both human and other animal studies to be compared more easily, and has led to mutual benefits with respect to both theory (Kraemer, 1992) and methods (Weaver and de Waal, 2002).

## TRENDS IN THE STUDY OF DEVELOPMENT

Our review of some developmental studies carried out since the publication of Tinbergen's (1963) paper has focused almost exclusively on studies asking causal questions using behavioral techniques. In recent years, many investigators have been attacking similar problems using neurobiological, endocrinological, and molecular biological methods and techniques that complement the purely behavioral studies. There have been neurobiological studies of filial imprinting (e.g., Horn, 1985, 2004), sexual imprinting (e.g., Bischof, 2003), and bird song learning (e.g., Nottebohm. 1991, 2000; deVoogd, 1994; Bolhuis *et al.*, 2000; Clayton, 2000; Bolhuis and Eda-Fujiwara, 2003; Maney *et al.*, 2003; Terpstra *et al.*, 2004). For example, neurobiological research into filial imprinting has concentrated initially on localizing the neural substrate for imprinting memory, and subsequently on analyzing the cellular and molecular mechanisms involved (Horn, 1985, 2004). Horn and his collaborators successfully identified a brain region (the intermediate and medial mesopallium) that is crucially involved in the learning process of imprinting and a likely site of memory storage. In birdsong learning, neurobiological investigations initially focused on the neural substrate of song production, leading to the identification of the "song system," a set of interconnected brain nuclei (Nottebohm, 1991; DeVoogd, 1994). In a recent review, Nottebohm (2000, p. 75) concluded that "We do not know, yet, where in [the ascending auditory pathway] reside the auditory memories that are eventually imitated" (see also Solis *et al.*, 2000; Bolhuis *et al.*, 2000; Bolhuis and Macphail, 2001). Only recently has the question of the neural substrate of tutor song memory been addressed explicitly (for review see Bolhuis and Eda-Fujiwara, 2003). It has become clear that the neural substrate for tutor song memory is likely to be found outside the "song system," in the caudo-medial nidopallium and/or the caudomedial mesopallium (Bolhuis *et al.*, 2000; Terpstra *et al.*, 2004). Another example of how these techniques are used to study developmental problems is the chapter by Crews and Groothuis in this volume. While molecular methods have been very useful in understanding many aspects of behavioral and neural development, it must be remembered that even a complete understanding of molecular mechanisms will not be sufficient for a complete understanding of behavioral development (Oppenheim, 2001).

Other investigators have asked functional and evolutionary questions about development. Oppenheim (1981) pointed out that stages in development are often not merely a kind of immature preparation for the adult state – although they can be – but that each developmental phase involves adaptations to the environment of the developing animal. He called these "ontogenetic adaptations." As a consequence, certain early behavior patterns may disappear in the course of development. One example is some of the movements a chick uses when hatching from its egg. Another is the finding by Hall and Williams (1983) that suckling is not a necessary antecedent for adult feeding in rats. An example of an initially causal investigation with important functional implications is Bateson's (1982) findings concerning sexual imprinting. He found that Japanese quail chicks (*Coturnix c. japonica*), reared in small sibling groups and then in social isolation preferred an unfamiliar individual when tested for social preferences when adult. Bateson suggested that such a choice could lead to optimal outbreeding. Ten Cate and Bateson (1988) later reviewed the literature and concluded that sexual imprinting could play an important role in sexual selection and the evolution of conspicuous plumage (see also ten Cate and Vos, 1999). The chapter by Crews and Groothuis in this volume gives other examples of how causal and functional thinking can complement each other in studies of development.

One trend that we have seen in our review is that the relationship between ethology and psychology has become closer since the publication of Tinbergen's paper. Studies of filial imprinting and bird song learning have been informed by concepts from the experimental psychology of learning, and studies of human social development (attachment theory) have been informed by concepts from studies of filial and sexual imprinting. The formation of neural representations is coming to play an ever more important role in conceptualizations of imprinting, song learning, and the development of behavior systems, and this parallels similar developments in cognitive psychology. A related trend is the increasing use of neural network models for understanding cognitive development in humans and other animals (Enquist and Ghirlanda, 2005). This cross fertilization has led to an overall advance in our understanding of developmental issues. We wish to conclude our chapter by highlighting some of these advances.

In many ways the nature/nurture debate has been settled by a realization that much of the problem was one of definition and/or specification of the question being asked (Lehrman, 1970). For example, we have not used the word "innate" in our discussion of development, but we have used the word prefunctional, which we define as developing without the influence of functional experience. (The word "innate" could be defined the same way, but it has so many other meanings that we, and most other students of development, prefer not to use it.) This does not mean that experience is not necessary for a behavioral mechanism to develop, but only that functional experience is not necessary. A simple example is the predisposition of young chicks to approach visual stimuli that have a certain resemblance to a conspecific: The predisposition requires non-visual experience before a visual preference can be expressed (Johnson *et al.*, 1989).

A related issue concerns the concept of modularity. We, and many others, use terms such as predispositions, templates, and behavior mechanisms. These concepts imply that particular parts of the central nervous system subserve particular functions, and that, by the time behaviorally interesting events are occurring, these parts, or modules, have a preassigned function. This means that, at the particular stage of development under consideration, the range of possibilities for further development of a particular behavior mechanism are so restricted that only special (i.e., already determined) kinds of experience can have a developmental effect on that mechanism. In practice, this means that, by the time of birth (or hatching), the central nervous system is already highly differentiated, with the general organization of pathways and connections already determined. By this stage of development, reversing the functions of major parts of the brain is generally impossible. Under these circumstances, it seems justified to speak of the song-recognition perceptual mechanism or the ground-scratching motor mechanism or the filial central mechanism as prefunctionally developed units of behavioral structure subject to further (but quite restricted) differentiation on the basis of subsequent experience.

It should be realized, however, that, if we follow the development of any behavior mechanism backward in time, we can always find a stage in which the nerve cells making up the behavior mechanism could have subserved a different behavior mechanism under somewhat different conditions. If we go back still further, we will find a stage when the cells could have become something other than nerve cells, and so on. At the time of birth – or any other arbitrarily chosen time – a particular set of nerve cells may have differentiated to the point where, if they survive, they will be the cells that mediate mate recognition, and in this sense, they are preassigned that function. But they are preassigned only from the point of view of future development.

There has been much recent discussion about the meaning of modularity with respect to higher cognitive functions (e.g., a language module or a spatial module). Some authors have suggested that modularity extends to the whole brain (Shettleworth, 1998; Cosmides and Tooby, 1994; Pinker, 1997), while others (Bolhuis and Macphail, 2001; Lefebvre and Bolhuis, 2003), following Fodor's (1983, 2000) original suggestions, have argued that modularity is restricted to input and output systems (perceptual and motor mechanisms in our terms). We will not comment on this issue further here except to say that the principles involved are the same as we have just discussed (cf. Karmiloff-Smith, 1992, 1998; for further discussion of modularity see Bolhuis, this volume and Lefebvre and Bolhuis, 2003).

Another issue we have discussed in this chapter is constraints on development: predispositions, sensitive periods, irreversibility. Here, too, the definition of terms has become clearer over the past 40 years, and these issues are no longer so controversial. Evidence from embryology and studies of the developing nervous system has put them into perspective (Oppenheim, 2001). Nonetheless, the debate over whether or not mallard duck females (Kruijt, 1985) or zebra finch males (ten Cate, 1989) have a predisposition or bias to prefer their own species shows that some of these issues can only be settled empirically.

SUMMARY AND CONCLUSIONS

Niko Tinbergen (1963) put behavioral development on the map as one of the four main problems in behavioral biology. Developmental research at the time was still in the grip of the nature/nurture debate. In his discussion, Tinbergen advocated an interactionist approach to development, which has been the main point of view in developmental research since then. In this chapter, we review research in a number of different areas, including imprinting, song learning, motivational systems, human language, and attachment. It has become clear that sensitive periods are important in all these areas, and that some aspects of development are generally irreversible; but these phenomena are not as rigid as was thought previously. Similarly, predispositions, or biases, play an important role in perceptual development, but here too there is much more flexibility than is sometimes suggested. Learning mechanisms underlying imprinting and song acquisition are very similar to those underlying other forms of associative learning. Cognitive concepts such as neural representations and behavior systems are becoming more common in the analysis of developmental questions. Finally, there are numerous parallels between developmental processes in humans and other animals.

# 6

# The study of function in behavioral ecology

Behavioral ecology is primarily concerned with the determination of the function, or survival value, of behavior (Krebs and Davies, 1993, p. 1). However, as should be evident from the fact that philosophers of science have devoted many volumes to debating the question of function, it is not straightforward even to define what "function" is, far less defend the validity of the various approaches used to investigate it. As it is relatively rare for behavioral ecologists, in common with most working biologists, to consider such seemingly rarefied debates, I start this chapter by examining the question of function. I place special emphasis on the development of Tinbergen's interest in the topic, but also try to draw conclusions in the light of recent treatments of the subject by philosophers and philosophically minded behavioral ecologists. I will then explore the reasons for behavioral ecologists' special interest in functional questions, and the merits of the different approaches to determining the function of behavior. Although the aims and methods of behavioral ecology are often portrayed as following the Tinbergian tradition within ethology, there are important differences. Finally, I will discuss where this has got us and, dangerous though it may be, I will attempt to forecast the future of the subject.

## TINBERGEN AND FUNCTION

When Niko Tinbergen was a student at Leiden University in the late 1920s, his notion of "function" in relation to animal behavior was probably that of any layperson with an interest in natural history: the role that behavior plays in the animal's life. The concept would not have had the intimate connection to evolution by natural selection that it does for modern-day behavioral ecologists, nor would Tinbergen have been worried about the teleological implications of the word.

*Tinbergen's Legacy: Function and Mechanism in Behavioral Biology*, Johan J. Bolhuis and Simon Verhulst (eds.). Published by Cambridge University Press. © Cambridge University Press 2008.

Although comparative and phylogenetic studies were pre-eminent in the universities of northern Europe at the time, an adaptationist stance was far from common. Indeed, van der Klaauw, the enlightened director of the Zoology Laboratory in Leiden, who promoted theoretical biology and encouraged Tinbergen as a young lecturer, said in 1940: "The prevailing theory of the last generation, Darwinism, has proven to be incorrect and one-sided" (van der Klaauw, 1940). As Tinbergen himself noted, he started his studies in Leiden "... at the tail end of a period of the most narrowminded, purely homology-hunting phase of comparative anatomy, taught by old professors" (Tinbergen, 1985). This is not to say that the young Tinbergen was not without Darwinian influences. Above all, Jan Verwey, who had been an assistant at the Zoological Laboratory when Tinbergen started his undergraduate studies, convinced Tinbergen that field ornithology and the detailed description of animal behavior were scientifically respectable (Röell, 2000). Verwey in turn, like several Dutch ornithologists of the time, had been influenced by Julian Huxley and Edmund Selous (Röell, 2000). Both had advocated the scientific value of studying animals in their natural habitat and both were ardent Darwinians. Huxley, particularly, had had a major influence on British ornithology, shifting its focus, at least partially, from taxonomy and distribution to the actual lives of the birds. Nearly four decades after his research career began, Tinbergen's (1963) "four Why questions" were to be based on Huxley's three divisions of biological investigation (Huxley, 1942), Tinbergen adding the question of ontogeny. That said, the writings of the young Tinbergen suggest he was more interested in careful description of behavior and in the fieldwork itself than in determining the adaptive or evolutionary explanations for it. Even his elegant, early field experiments on the homing of the beewolf (Tinbergen and Kruyt, 1938), were directed to understanding causal factors, not function, and drew their acknowledged inspiration from earlier work by Von Frisch and his students (Röell, 2000, pp. 69–76).

Meeting Konrad Lorenz in 1936 was clearly a turning point. Lorenz is maybe seen today largely as a (successful and important) publicist for ethology as a discipline, and as the author of now superseded theories of causation (for example, the "hydraulic model" of motivation; the limitations of Lorenz's model are standard textbook material, but see Slater, 1999, for a recent account). However, there is clear evidence (see Röell, 2000, pp. 111–121) that it was Lorenz who in fact convinced Tinbergen not only that the causal analysis of behavior should be studied within a firm theoretical framework but that behavior should be studied in a phylogenetic and evolutionary context. Once he left for Oxford at the end of the 1940s, just a couple of years after being offered the Chair in Experimental Zoology at Leiden, he became immersed in a culture where evolution by natural selection reigned supreme; for this was the Oxford of E.B. Ford, Arthur Cain, and David Lack. Functional and evolutionary analysis of behavior now started to take center stage in Tinbergen's research. Furthermore, the great international success of Tinbergen's landmark (1951) book *The Study of Instinct* gave him invitations to travel to the United States, where he was exposed to leading evolutionary thinkers such as Ernst Mayr, who subsequently devised his own, still influential, framework for the multilevel analysis of organisms, the "proximate-ultimate" dichotomy (Mayr, 1961). And so the theoretical framework we now recognize as "The Four Whys" crystallized in Tinbergen's mind. In addition to causal and ontogenetic explanations, one must deduce the survival value of behavior and its evolutionary origins.

### MULTIPLE FUNCTIONS

The term 'function' is used in many different ways, even within biology (see e.g., Bock and von Wahlert, 1965; Hinde, 1975; Wouters, 2003) but, within behavioral ecology, two are particularly important. First, as synonymous with current survival value or, more generally, because survival is only one component of fitness, function is defined as the means by which a behavior increases fitness. This is an extension of the everyday use of "function" (What is it for? What is the role of the behavior in the animal's life?), but with "for" and "role" assessed in terms of current fitness. We can call this a "current utility" definition of function, as it concerns the current value of the behavior in promoting gene replication. Most would see this as equivalent to the means by which natural selection maintains the genes for the trait in the current population (relative to alleles for alternative traits, or the complete absence of the trait), although we return to this point later on. Hinde (1982) distinguishes between this form of function in the weak sense (any beneficial consequence) and function in the strong sense (those beneficial consequences through which natural selection acts to maintain the trait). However most behavioral ecologists today would accept that any beneficial consequence contributes towards the maintenance of the trait and it may be artificial to distinguish between weak and strong functions.

The second use of the term function concerns the selective advantage of the behavior that led to its current prevalence. That is,

$y$ is the function of $x$ if, and only if, $x$ became prevalent through its advantage in performing y (modified from the definition of an adaptation in Sober, 1984). The distinction between function in the historical sense and function as current utility precisely mirrors Gould and Vrba's (1982) distinction between an adaptation and an aptation. In their terminology, we should call traits that increase current fitness "aptations" and reserve the term 'adaptation' only for aptations that have evolved because of the same selection pressures that currently maintain them. Gould and Vrba would therefore say that historical function refers to adaptations and utilitarian function to aptations. The terminology highlights a simple fact even a committed pan-adaptationist, with no time for phylogenetic or genetic constraints, would accept: current function may not equal past function. This is potentially a serious problem for behavioral ecologists interested in establishing historical function; investigating current utility may not tell you why the trait evolved in the first place. For example, take the question "What is the function of ultraviolet vision in birds?" Utilitarian functions in mate choice and foraging have been clearly established by experimentation (for review see Cuthill *et al.*, 2000). However, one's faith that ultra-short-wave (SW1 or ultraviolet/violet) visual pigments evolved because they help in detecting the urine trails of voles (Viitala *et al.*, 1995), or assessing plumage coloration (Bennett *et al.*, 1997; Hunt *et al.*, 1999), might be shaken by the knowledge that marine turtles have very similar visual systems to birds (Ventura *et al.*, 2001; Zana *et al.*, 2001). Indeed, the fact that SW1 visual pigments and tetrachromatic color vision originated very early in vertebrate evolution (Bowmaker, 1998; Yokoyama and Shi, 2000), makes it clear that the historical origins of ultraviolet vision lie in selection pressures acting within a shallow aquatic environment.

The example above is deliberately extreme in order to expose the reasons why some behavioral ecologists and other adaptationists can get quite heated at the suggestion that the adaptations they study should be demoted to "mere" aptations. First, it suggests that studies of present day organisms, and the current utility of behavior, cannot expose historical function; only methods that probe deep history are suitable. As stated, this seems a valid point but, taken in the light of what behavioral ecologists actually do, as we shall see later, it is false. Second, it suggests that establishing that a trait evolved in ancestors for a (historical) function different from its present use is sufficient to explain its existence in the living descendants. It is not, as one also needs to establish what has maintained the trait to the present day (see e.g., Reeve and Sherman, 2001). For example, having established the deep historical function that gave rise to ultraviolet visual pigments in Cambrian fish, one still has to explain why the trait has been retained in members of groups with as diverse ecologies as teleost fish, reptiles and birds. The opportunities of 500 My of evolution, and the fact that visual pigments can be lost, or their spectral tuning modified by a single amino acid substitution (Yokoyama, 1999), renders arguments of phylogenetic "inertia" or constraint untenable. Therefore, even if ultraviolet-sensitive visual pigments evolved in Cambrian fish for reasons unrelated to the challenges faced by present-day birds, it is perfectly legitimate to ask "What is the function of ultraviolet vision in birds?" This suggests that a proper definition of historical function should extend beyond the reasons behind the origin of the trait, and should extend right up to the present day to include the explanation for its current maintenance. Some authors go further. For example, Amundson (1996) and Reeve and Sherman (2001) argue that adherence to a historical definition of function inevitably leads to confusion, due to the mixing of evidence suitable for establishing phylogenetic history and for current phenotype adaptedness. Reeve and Sherman (2001) therefore argue strongly for a pure, current-utility definition of function and, in so doing, argue for a stronger distinction of Tinbergen's "evolution" and "survival value" questions than most behavioral ecologists currently make. Before considering this position, and the extent to which it matches Tinbergen's own position, we need to consider what philosophers of science have said on the issue of function.

### PHILOSOPHY AND FUNCTION

Philosophers have rightly been wary of functional approaches in biology, because to ask the question "What is behavior '$x$' for?" is teleological; it implies purpose. Whilst a valid question for an artefact created by an intelligent designer, of which *Homo sapiens* and perhaps a few other animals are the only known examples, how can it be a valid question for a trait that has evolved through the blind mechanism of natural selection? The harsh answer is that it is not a valid question: bird wings, bat sonar, meiosis and the human brain have no purpose, they simply appear to have a purpose because of the effects of natural selection in the past. They are not "designed," they are "designoid" (appearing to have been designed; Dawkins, 1996). That said, there seems to be a general consensus amongst philosophers of science that it is, in fact, valid to give functional explanations in biology, as long as one remembers that saying "*x* is for doing *y*" is shorthand for "*x* appears to serve this purpose in environment *y* because of past relationships between variants of *x* and fitness in environment *y*." This linguistic shortcut is more than just convenient; the "adaptationist stance" of asking design questions about nature helps frame questions about mechanism in a directed and efficient manner (Dennett, 1987, 1995). This is true outside of biology. For example, there was a sea-change in the study of visual perception when David Marr distinguished questions of computation (what is computed and why), from algorithm (how to achieve the desired computation) and implementation (physical realization via silicon or nervous tissue) (Marr, 1982). Marr's computational question, which contains a strong functional component, helps direct the mechanistic questions of algorithm and implementation. Ask an engineer how a mystery machine works and their first question will be "What is it for?" (see Dawkins, 1996; Dennett, 1987, 1995).

This would seem to suggest that the historical definition of function should take precedence, and even that the utilitarian definition is empty. Indeed, more philosophers of science veer towards this view than not (Brandon, 1990; Amundsen, 1996), yet amongst biologists the reverse appears to be true: the current-utility definition of function predominates. Sometimes this is an ideological stance argued from logical principles (Thornhill, 1990; Reeve and Sherman, 2001); more often it is a passive reflection of the status quo. Current fitness effects are easier to study and are what the Tinbergian tools of field observation and experimentation have equipped behavioral ecologists to do. By defining function as current utility we can dodge the tricky issue of whether what we are studying really tells us about a trait's adaptive origins, that is, function in the historical sense. I believe a determination of utilitarian function can tell us (a lot) about historical function, and argue this in the next section in the context of different approaches to examining function. However, perhaps perversely, as a biologist who agrees strongly with the views expressed by Thornhill (1990) and Reeve and Sherman (2001), I side with the philosophers in accepting that function, if its everyday teleological meaning is to make any sense, should have a historical definition. Function defined as an explanation for the current maintenance of the trait is not in fact synonymous with function defined as current fitness effects, because "maintenance" implies historical continuity, even if it is only very recent history. For "function" to retain any vestige of its common meaning "What is it for?", then it needs at least a shallow historical component. Of course, the best way to establish how a trait is maintained by selection is to study current fitness effects of the trait compared to alternatives (Thornhill, 1990; Reeve and Sherman, 2001). Furthermore, it is important to separate accounts of phylogenetic history from phenotypic existence (Reeve and Sherman, 2001). It is just that it may be ill advised to label the latter "function" and strip the word of any historical meaning.

Even replacing "function" with the apparently more neutral term that Tinbergen favored, "survival value," does not remove the problem. Certainly one could say that there is an evaluation in the currency of fitness, but why use the term "value" at all, why not just "fitness effects?" Other than the scientist measuring fitness, there is no evaluator; there are just effects on fitness due to (natural or experimental) variation in the trait. This is in fact what Tinbergen meant by function and for this reason it is wrong to see Tinbergen's functional and evolutionary "why" questions as subdivisions of Mayr's "ultimate" question (Dewsbury, 1999; Wouters, 2005). Mayr (1961) was dealing with explanations for biological phenomena, and he made the point that traits can be explained as the result of effects taking place within the lifetime of the individual (proximate explanation) or over evolutionary time (ultimate explanation). Tinbergen's "survival value" concerned fitness effects, and effects cannot be causes, so function defined as current utility cannot be an explanation for a behavior. For function to have explanatory status, to be a cause not an effect, it has to have at least a shallow historical component such as maintenance through the action of selection in recent generations.

## BEHAVIORAL ECOLOGY AND FUNCTION

Studying historical function is difficult, most notably because you cannot go back in time and do the necessary experiments. The environment has changed since the trait evolved and, even if one could recreate something close to the ancestral environment, the animal has changed. So the pragmatic solution would seem to be to study current utility, cite Tinbergen (1963) and, if challenged, say one never intended to make any claims about function, in the historical sense, as an ultimate explanation for the behavior. In the light of philosophers' debunking of current utility function as an explanatory account of *anything*, is this a reckless thing to do? The answer is that it depends on what sort of biologist you are.

Many of the first behavioral ecologists were ecologists, which is clear from the fact that "behavioral" is the adjective not the noun. The origins of Optimal Foraging Theory lay, in large part, in attempts to predict population and community processes rather than behavior per se (e.g., Emlen, 1966; MacArthur and Pianka, 1966). The optimal diet model made a simple assumption about predator behavior in order to predict diet breadth and community stability; Ideal Free Theory was devised to explain habitat choice on a scale larger than individual foraging patches; one of the first derivations of the Marginal Value Theorem was with reference to habitat emigration thresholds (for all examples, see Stephens and Krebs, 1986). If one has these motivations, it is the current effect of behavior that matters rather than its historical origins. Similarly, if interest in the behavioral ecology of a species lies in predicting changes in population size, social or genetic structure (i.e. for conservation purposes), then it is the current fitness effects of behavior that matter. Although these are all legitimately classed as behavioral ecology (for example, see the remit of the journal Behavioral Ecology), the fact is that most behavioral ecology is concerned with explaining the design, or adaptedness, of behavior. To do this requires the determination of function in the historical sense (both shallow and deep history). The different methods used by behavioral ecologists allow this to be inferred with different degrees of confidence, so it is important to consider each in turn.

Figure 6.1a characterizes the classic Tinbergian approach to investigating behavior. The inspiration for an investigation nearly always came from detailed observations of an animal in the field, although it sometimes came from recognition of a cross-species association between particular traits and particular ecologies. The observations led to a functional hypothesis that was used to generate predictions that could be tested experimentally. It is the latter, the elegant use of simple field experiments, that Tinbergen is rightly credited with taking to new heights. Modern behavioral ecologists usually start in the library or at the computer rather than in the field, because the driver for most contemporary behavioral ecology is the testing of theory (Fig. 6.1b), usually a particular mathematic model. This is understandable, because the theory is so much more developed than it was in Tinbergen's day. Although the roots of many of the core ideas in behavioral ecology already existed (e.g., sexual selection and life-history theory; Fisher, 1930), they were not mainstream, and other key theoretical developments (e.g., kin selection, evolutionarily stable strategies) arrived only very late in Tinbergen's career. However, it is somewhat strange that Tinbergen did not make use of two of the applications of economic theory that are the theoretical mainstays of behavioral ecology:

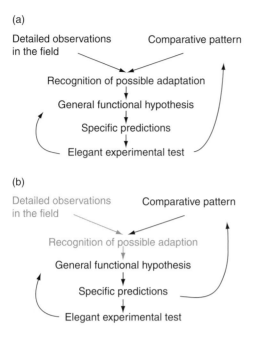

(a)

Detailed observations    Comparative pattern
in the field

Recognition of possible adaptation

General functional hypothesis

Specific predictions

Elegant experimental test

(b)

Detailed observations    Comparative pattern
in the field

Recognition of possible adaption

General functional hypothesis

Specific predictions

Elegant experimental test

Fig. 6.1. (a) A caricature of the sequence of events in an investigation by Tinbergen. The initial stimulus was usually an observation of behavior by a wild animal in its natural state. (b) The equivalent flow chart for contemporary behavioral ecologists. The initial stimulus is usually the desire to test a particular theory.

optimality and game theory. After all, in 1969, Niko's brother Jan won the Bank of Sweden Prize in Economic Sciences in memory of Alfred Nobel (often loosely referred to as the Nobel Prize in Economics), and was an expert on utility theory. Instead, the economic approach to behavior was introduced via (i) the early optimality models of, amongst others, Pulliam (1974) and Schoener (1971); and (ii) the first applications of game theory (Maynard Smith and Price, 1973). For Tinbergen, the primary tools were observational and experimental fieldwork.

## Observational studies

Observation in the field was Tinbergen's main source of inspiration for research, not just because he was first and foremost a naturalist, but also because he firmly believed that to understand function you have to appreciate the context in which the behavior is performed. This remains true today, even though behavioral ecology is a theory-driven discipline. An hour spent watching an animal behaving naturally can generate more interesting questions about behavior than a day spent in the library. Furthermore, simple observation has the potential to exclude particular functional hypotheses. For example, the hypothesis that a certain male display is for mate attraction would be rejected, to the satisfaction of most scientists, by the observation that the male only performs the display outside the breeding season and when no females are present. That said, purely observational studies can rarely provide a totally convincing test of a hypothesis, because the possibility of confounding variables always exists. So, for example, if one uses natural variation between individuals to test the hypothesis that a particular sexual ornament functions to signal male quality to females, then any correlation between ornament size and male quality, as predicted by this hypothesis, may in fact be due to any of the other factors that vary between males. Even if one controls for likely other factors (e.g., body size, condition, age, experience) statistically, or through sample selection, then an infinite number of possible other factors remain. Of course, a study that convincingly demonstrated a strong effect of the factor of interest, and no influence of all the likely confounding variables, would be good circumstantial evidence for the hypothesis, but it would not be conclusive.

The type of function that observational studies have greatest bearing on is current utility, and the most informative studies are those which measure currencies closest to fitness. Hence, a study relating a bird species' clutch size to offspring recruitment and both parental survival and future fecundity is a more powerful test of the hypothesis that clutch size is optimized than a study that measures fledging success and parental body mass. Measurement of selection coefficients and genetic correlations between different characters are particularly informative about current function precisely because they are near-fitness measures. However, it is important to realize that, however sophisticated the techniques used, they suffer from the same limitations of possible confounding variables as any non-experimental study. Furthermore, the fact that they are informative about possible evolutionary trajectories in the short term (e.g., the rate of response of trait A to selection is constrained by opposing selection on correlated trait B) does not mean that the same constraints apply in the long term (Hammerstein, 1996). Measurements of the response to selection only estimate the constraints on selection with reference to the current genetic influences on the trait and under the prevailing genetic architecture. This undoubtedly constrains selection in the short term (a developmental or genetic constraint, as discussed under *The Comparative Method*), so is a vital consideration for those biologists interested in the consequences of immediate selection. However, in the longer term, alternative genetic influences (with different genetic constraints) may evolve and the genetic architecture can itself evolve. Extrapolating backwards, rather than forwards, from the present day, this suggests that estimates of the constraints on adaptation today may not accurately represent the constraints on adaptation at the time of initial selection for the trait (Hammerstein, 1996). In short, a consideration of the constraints affecting current utility may not reveal the constraints that influenced historical function. Whether the establishment of current utility tells one anything at all about historical function is considered in the next section on experimental manipulation.

### Experimental manipulation

The primary tool of the behavioral ecologist, and a direct legacy of Tinbergen, is the phenotype manipulation experiment. Thus, just as Tinbergen used dummy eggs to investigate the function of egg shell removal by gulls (Tinbergen *et al.*, 1962), modern behavioral ecologists alter tail length, coloration, parasite load, clutch size, you name it (see Krebs and Davies, 1993 for countless examples). Done properly, this approach has a unique power to establish function in the sense of current utility, as it directly relates the trait to its fitness consequences. No other approach, apart from the optimization model (see later), is as useful in determining current function (see Reeve and Sherman, 2001, for a detailed advocacy). Indeed, the most powerful approach of all, as widely used within behavioral ecology, is to combine predictions from optimization models with experimental tests. The goal is to relate experimental variation in a trait to changes in fitness measures predicted by the functional hypothesis. An alternative experimental approach, borrowed from evolutionary biology but increasingly used within behavioral ecology, is the selection experiment (e.g., Reznick *et al.*, 1990; Wilkinson and Reillo, 1994). This is, in many ways, the complementary approach, in that experimentally induced changes in fitness (through artificial selection) are related to predicted changes in the trait of interest. In either case, the mark of a good experiment is one in which the manipulation changes one, and only one, factor compared to a control group or groups. If this is achieved (and it is trickier in practice than in principle) then one can conclude that there is positive evidence for the proposed (current) function. It is worth noting here that field experiments are particularly informative in this regard. They are often technically more difficult than lab experiments, and are liable to have lower power to detect the predicted effects because of the uncontrolled variation against which those effects must be assessed. However, this is traded off against greater external validity (Mook, 1983), namely the range of conditions over which the results generalize. As it is the animal's natural environment that is the context in which a function applies, determination of fitness effects in that context will be the most persuasive evidence for the proposed function. However, one cannot, on this evidence alone, rule out other functions that may be as, or more, important. If the predictions of the proposed hypothesis are unique, or they are sufficiently numerically precise to render a chance fit to the experimental results statistically unlikely, then faith in the pre-eminence of the proposed function is increased correspondingly (Kacelnik and Cuthill, 1987).

If phenotype manipulation and selection experiments provide strong evidence for function in the current utility sense, what can they tell us about historical function? As they do not estimate fitness effects under the exact environmental conditions at the time the trait initially evolved (including the state of all organisms interacting with the study organism), and it is possible the genetic background of the organism has changed, one might conclude "not very much." However, an estimate of current function provides the best working hypothesis for historical function; it establishes that the hypothesized historical function is plausible and can be important. In as far as "adaptedness" (a close fit between an organism's design and its environment and lifestyle) was recognized by pre-Darwinians from Cicero (106–43BC) to Paley (1743–1805) (Dawkins, 1986), it seems that historical selection pressures usually match current ones. There is no logical necessity that they should but, if they did not, then widespread apparent design would never have been recognized. This gives the adaptationist hope that an examination of current fitness effects does in fact shed light on historical function.

## Modeling

Behavioral ecologists principally use economic models based on phenotype evolution unconstrained by the details of genetic mechanisms (Grafen, 1984), but also explicit genetic models. This is not the place to examine the merits of different types of modeling approaches (see

Grafen, 1984; Hammerstein, 1996; Abrams, 2001), but rather to focus on what such modeling tells us. In the case of either phenotypic or genetic approaches, the model is the primary tool for investigating "good design" in behavior. Certainly, an important role of modeling is to generate predictions for experimental tests (see below) but, more importantly, it exposes the logic of an adaptationist argument. Does behavior $x$ maximize fitness in environment $y$, given the assumed constraints, or not? Is strategy S evolutionarily stable given particular starting conditions? Whilst in principle it may be possible to answer such questions experimentally, in practice only mathematics can provide an unambiguous, general solution. Furthermore, only mathematics can establish the optimal behavior in environments and organisms that no longer exist, or might exist in the future (e.g., predicting the effects of global warming). As such, mathematical modeling can provide strong evidence for function in both the current utility and the historical sense. This comes with an important proviso, namely that plausible alternative hypotheses can be rejected. Showing that behavior $x$ maximizes fitness in environment $y$, given constraints a, b and c, is a necessary condition for accepting this particular functional hypothesis, but it is not sufficient. One must show that behavior $x$ does not maximize fitness under plausible alternative constraints, or plausible alternative characterizations of the environment. For example, modeling indicates that the trichromatic color vision of the bee, based on ultraviolet, "blue" and "green" visual pigments, is optimal for discrimination of floral color signals (Chittka and Menzel, 1992). However, no matter how obviously plausible this hypothesis is, one cannot conclude that floral discrimination is the historical or current function of the precise spectral tuning of bee visual pigments until plausible alternative functions are shown to predict different visual pigment sensitivities from those observed. In fact, the historical origins of these visual pigments must lie in other functions, because this form of trichromacy is typical for all insects and long pre-dates both bees and flowers (Chittka, 1996). The ubiquity of bee-like color vision across insects with a vast diversity of ecologies leads Chittka and Briscoe (2001) to conclude that bee vision is not explained by adaptation but by phylogenetic constraint. However, the need to test and reject alternative models also applies to the acceptance of non-adaptive hypotheses, just as it does to the uncritical acceptance of adaptive hypotheses. First, bee vision may well be maintained by selection for floral discrimination (current utility and, indeed, a historical function going back to the ancestors of bees), but it just so happens that the (different) visual tasks faced by parasitoid wasps, beetles and crickets also favor the same spectral tuning. This may seem unlikely, and it is this that leads Chittka and Briscoe to their non-adaptive explanation; however, these alternative models have not been constructed, so we simply do not know. Second, it could be that an alternative visual task to flower discrimination (e.g. discriminating sky from leaves or leaves from earth, or navigation) selects for bee-like trichromacy in both bees and other insects, both past and present. Again, the relevant models have not been constructed or tested. Third, cause and effect might be the reverse of the original hypothesis: flowers may have evolved so that their colors are readily detected or discriminated by insects (be they bees or other pollinators). So, a good fit between a functional model and observation should be treated with suspicion before alternatives have been rejected; but also, the ubiquity of a trait in the face of no obvious ubiquitous selective force should not be treated as evidence for lack of an adaptive explanation.

### The comparative method

The discussion above leads naturally to the comparative method and what information it can and cannot provide. The origins of what is now known as the Comparative Method (in fact a suite of methods, each having their own advocates) do not lie in evolutionary biology, but in ethology and ecology. Perhaps surprisingly, it is only in the last two decades that phylogenetic reconstruction has become a central part of comparative behavioral ecology. Although inter-specific comparison was an approach used by Tinbergen and his students to study adaptation (Cullen, 1957), most observers trace the advent of the quantitative comparative method to Crook's (1964) study of weaver bird social organization. The logic is that if, across a set of species, a given trait is consistently associated with a given environmental or social factor, then one might infer that the trait is an adaptation for that factor. The problem with Crook's approach was that closely related species are not statistically independent, because they share many common features (ancestry, ecology) as well as putative adaptation to the proposed selective agent. Crook himself did not see this as a problem for the study of social organization in relation to ecology. Indeed, one of his agendas was to break down the Lorenzian conception of behavior as species-specific and phylogenetically conserved, and instead launch a new discipline of social ecology in which adaptive phenotypic plasticity was the focal concern (Burghardt and Gittleman, 1990). Because his studies of primates, in particular, revealed considerable plasticity even within species, and the intra-specific variation mirrored the inter-specific variation, it seemed reasonable to ignore phylogenetic non-independence of species (these were not Crook's words, but summarize the research strategy; Burghardt and Gittleman, 1990). However, this may not be true in general and so the general statistical problem remains for comparative tests of adaptation. After proposed solutions involving analysis at higher taxonomic levels than the species, the field has settled on statistical methods that "control for" phylogeny, of which the first and still most widely used was Felsenstein's (1985) method of phylogenetically independent contrasts (for reviews of the different approaches, see Burghardt and Gittleman, 1990; Harvey and Pagel, 1991; Martins, 1996; Harvey and Nee, 1997; Freckleton et al., 2002). Although recent, the approach is now so widely used that the modern comparative method is seen as one of the key tools in behavioral ecology (Harvey and Nee, 1997).

The unification of comparative methods used for inferring adaptation in living organisms and for reconstructing the phylogenetic history that gave rise to this variation, would seem to be a very positive step forward (Harvey and Nee, 1997). However, whilst partitioning interspecific variation in a trait into components due to phylogeny and components due to a specific adaptation is statistically sound, there can be considerable misunderstanding as to what the variance "due to" phylogeny represents. What it should be taken to represent is all the shared features of the species other than the ones that comprise the adaptive hypothesis under test (see e.g., Blomberg and Garland, 2002; Freckleton et al., 2002). As closely related species inevitably share many features, it is a real possibility that one or more of these (unmeasured) features is actually the selective force responsible for the interspecific variation observed, and not the hypothesis under test. Therefore, as in any statistical analysis, one should attempt to control for these possibly confounding variables. What the variance "due to" phylogeny should not be taken to represent is evidence of an inability to adapt (which is that the terms phylogenetic constraint or inertia imply), or as an alternative explanation to that of adaptation through natural selection (see Orzack and Sober, 2001; Reeve and Sherman, 2001; Blomberg and Garland, 2002). Lack of the appropriate genetic variation, genetic drift, or a rapidly changing environment may indeed constrain adaptation in some situations, but phylogenetic analysis does not provide evidence for it. Phylogenetic conservatism within a clade (as in the insect visual pigments discussed previously) may certainly limit the power of the comparative method to test adaptive hypotheses, but if one wants to advance a non-adaptive alternative hypothesis, such as a developmental or genetic constraint, then one needs direct evidence to support it.

With a philosophical position that defines function as current utility, and an understandable antipathy towards claims for the pre-eminence of phylogenetic methods, Reeve and Sherman (2001) downplay the importance of the comparative method for testing hypotheses about adaptation. Yet repeated independent and convergent evolution of a trait (the cladist touchstone for adaptation) that coincides with the distribution of a particular hypothesized selective force, is very powerful evidence for function both historical and current. It is not conclusive evidence because, no matter how rigorously possible confounding variables are controlled for, or alternative hypotheses tested, the method is ultimately correlational. Establishment of correlation does not, of course, prove causation. Failure to detect the hypothesized association may be even less informative. If the non-significant association is because of phylogenetic conservatism of the trait(s), this is effectively a failure through lack of statistical power. This is no ground for rejecting the functional hypothesis under test (as Reeve and Sherman, 2001, correctly assert); but neither is it a reason to reject the comparative method in general.

BEHAVIORAL ECOLOGY PAST, PRESENT AND FUTURE

In the textbook that, to many, defines the compass of the field, Krebs and Davies (1993) portray behavioral ecology as part of the Tinbergen legacy: the four whys, the emphasis on behavior observed in its natural context, the simple, yet elegant field experiments. To what extent is this true? To what extent has it succeeded as a discipline? And what does the future hold?

Whilst it is easy to see behavioral ecology as ethology minus the proximate questions, or fashionably renamed to avoid association with dated models of causation, it is not. Wilson (1975, p. 5) portrayed it as a separate discipline, effectively synonymous with his "sociobiology," but contemporaneous with ethology even in 1950. Certainly behavioral ecology and sociobiology drew from ethology the Tinbergian influences mentioned above, but its theoretical foundations come from ecology, evolutionary biology (sexual selection, kin selection, the gene's-eye viewpoint, genetic conflict) and economics (optimality and game theory); the comparative method came from Crook's social ecology, not Lorenz's studies of behavioral phylogeny (e.g., Lorenz, 1950). Indeed, the behavioral ecology and sociobiology of Wilson in 1975 was primarily geared towards explaining population biology and social organization, using evolutionary theory as a tool, rather than explaining the evolution of behavior per se (Wilson, 1975, pp. 4–5). That is, Wilson was more concerned with functional questions of the current utility type rather than historical function, because it is the former that affect current population processes. As I said earlier, the pioneers of Optimal Foraging Theory had similar aims, being ecologists not ethologists, but this is not the way the subject has developed since 1975. Wilson (1975; figs 1, 2) predicted that, by the year 2000, behavioral ecology's most important interaction would be with population biology, but that has (sadly) not happened. The dominant question asked by behavioral ecologists is about design.

Because of the interest in design, and the replicator-fitness-centered approach, the most closely allied field to behavioral ecology is probably evolutionary biology. However, although design is the result of evolution by natural selection, behavioral ecology is not simply a subdiscipline of evolutionary biology. Behavioral ecologists increasingly ask questions about mechanism and development, because (i) these determine some of the constraints within which behavior must be optimized (Stephens and Krebs, 1986), (ii) mechanisms may hold clues to the trade-offs between different fitness components (e.g., How does the immune system mediate the link between mating effort and disease resistance, if it does?; Westneat and Birkhead, 1998) and (iii) because one can only reliably expect the mechanism to be adaptive within the context within which it evolved, an understanding of the mechanism may explain why behavior can be maladaptive when placed in a novel context such as the laboratory (Houston and McNamara, 1999; Steer and Cuthill, 2003). Also, I would hope that behavioral ecologists now ask mechanistic questions because, like ethologists, they seek to understand behavior at all levels of explanation. Because of this new interest in mechanisms, behavioral ecologists have sought out collaborations with disciplines that neither Wilson, nor Tinbergen, could have predicted, such as parasitology, immunology, and endocrinology. Links to neuroscience were predicted by Wilson, but he saw them as mediated via a common link to ethology and physiological psychology, whereas in many cases it is direct (e.g., interest in how sensory systems shape signal design; Endler and Basolo, 1998; Autumn *et al.*, 2002). Because of their special focus on function, and the light that an understanding of function sheds on why a mechanism operates the way it does, behavioral ecologists are in a strong position to provide the stimulus for new interdisciplinary links. Whilst obviously not the only biologists interested in design (e.g., functional morphologists, sensory ecologists, perceptual psychologists), these other "design sciences" traditionally look to physics and engineering for their mathematical theory. As a result the optimization criteria are things such as maximizing signal: noise ratio, energy efficiency or detection distance (Cuthill, 2002). However, natural selection does not operate in S.I. units, it maximizes fitness. Behavioral ecologists, having originally used currencies relatively remote from fitness (e.g., net energetic intake rate while foraging), these days are careful to make quite explicit links to true or near-fitness currencies. Therefore I feel quite optimistic that behavioral ecologists can contribute a lot to these sciences, as well as benefiting from the knowledge they provide on mechanisms.

To what extent do behavioral ecologists deliver on their promise of explaining function? There are certainly critics of behavioral ecology as a whole (the most famous being Gould and Lewontin, 1979) and specific components of the research program, such as optimality theory (e.g., Pierce and Ollason, 1987). What these attacks fail to do is suggest plausible alternative methods for determining function. So yes, determining historical function may be difficult and yes, there may be genetic or phylogenetic factors that constrain adaptation, but what exactly would one do differently? Optimality modeling is specifically designed to investigate the effect that different assumptions about constraints have on the strategies that maximize fitness. Phenotype manipulation experiments are the way to assess whether a trait affects fitness in the proposed way; selection experiments can reject the hypothesis that a trait is constrained from evolving towards a predicted optimum. Cross-species comparison (in the Darwinian sense, rather than the formal comparative method) indicates that particular phenotypes are attainable by natural selection. So, the approaches taken by behavioral ecologists are exactly the ones that should be used to expose and understand constraints on adaptation. Certainly there are some things that can be done differently by behavioral ecologists, such as to be more rigorous in testing alternative functional hypotheses and designing experiments with the power to reject rather than simply fail to confirm. There is evidence that many studies in behavioral ecology have only moderate statistical power at best (Jennions and Møller, 2003), so there is a real risk of a bias in the literature towards positive results. However, this is not a criticism specific to behavioral ecology but rather what is expected in a young discipline which is expanding; the impetuousness of youth, if you like. Most behavioral ecologists would accept that the "Spandrels" critique and its ilk served a useful role in forcing self-assessment and sharpening the field both theoretically and methodologically. What behavioral ecologists firmly reject are the real alternatives to the adaptationist research programme, which are either a denial that natural selection has produced and continues to shape "good design" in organisms, or a research program driven only by description of the data or mechanisms. A science of behavior that concerns only pattern and/or has a view of process that fails to address why the process (mechanism) exists, is incomplete (and, to me, extremely boring).

Would Niko Tinbergen recognize modern behavioral ecology? He would value the sharpening of theory and incorporation of more rigorous modeling approaches from economics. He would value the sophistication of modern comparative methods. He would recognize his approach in almost all the experiments published within the discipline. However, he would lament the decline in the number of students trained in the careful observation and description of wild animals. Modern behavioral ecologists are so keen to test theories that the rich details of behavior are often lost. There is an element of the behaviorism of B.F. Skinner in adoption of experimental paradigms that provide reliable data, at the expense of all else. Tinbergen would hate that. However he would undoubtedly recognize the relationship between his ethology, and behavioral ecology today. It is a relationship of descent with modification, but involving considerable recombination with other disciplines. The new ideas and techniques that this has brought have invigorated the field and proved an effective defence against culture's equivalent of Muller's ratchet: the accumulation of stale ideas. They have also allowed the field to keep pace with the Red Queens of other disciplines in competition for limited grants, journal space and the minds of the next generation. If behavioral ecology is to have an equally high validity in 25 years as it has now, then one thing is certain, it will have to keep evolving.

ACKNOWLEDGMENTS

I am very grateful to Johan Bolhuis, on behalf of the Royal Dutch Zoological Society, the Dutch Society for Behavioral Biology and the Dutch Foundation for Scientific Research, for inviting me to talk at the symposium in Leiden that led to this chapter. It was an honor to be involved in the meeting to celebrate Niko Tinbergen's landmark paper, but also great fun to think about our science rather than just doing it. I am particularly grateful to Kate Lessells and Marcel Visser for their great hospitality, as ever, at the Netherlands Institute of Ecology, Heteren. Thanks also to Arno Wouters for putting me right on what philosophers think and for access to unpublished manuscripts, to D.R. Röell for writing the fascinating book that provided my information on early Dutch ethology, to Simon Verhulst for comments on the manuscript, and to Alasdair Houston for much coffee and discussion.

MICHAEL J. RYAN

## 7

# The evolution of behavior, and integrating it towards a complete and correct understanding of behavioral biology

### EVOLUTION AND BEHAVIOR

In his classical discourse on aims and methods in ethology, Niko Tinbergen (Tinbergen, 1963) first posed a single, central (and what Tinbergen referred to as "admittedly vague") question: *Why do animals behave the way that they do?* (p. 411). He suggested four aims, questions, approaches, or levels of analysis that can be used to address this question: causation, survival value, ontogeny, and evolution. We are here to celebrate the anniversary of that paper that was of such important heuristic value for our field. My purpose is to consider aspects of the evolution of behavior.

In this paper, Tinbergen stresses more than once that the large question in ethology, mentioned above, is a question of the biology of behavior, and he gives praise to Konrad Lorenz's insistence in stressing this notion. It is in this context that he argues for an integrative analysis of behavior that addresses several important aspects of its biology: the physiological mechanisms regulating the behavior, the current adaptive significance of the behavior, the acquisition of the behavior by the individual, and the past evolutionary history of the behavior.

### The virtues of integration

There is no doubt that Tinbergen appreciated that an integrative approach to animal behavior would result in a more complete understanding of the main question that motives us – why do animals behave as they do? Consider, for example, how much more incomplete our

*Tinbergen's Legacy: Function and Mechanism in Behavioral Biology*, Johan J. Bolhuis and Simon Verhulst (eds.). Published by Cambridge University Press. © Cambridge University Press 2008.

understanding, and perhaps our appreciation, of the vocal acrobatics of song birds would be if we did not understand how that richness in repertoire was derived by the neural mechanisms in the sound control nuclei in the brain (Nottebohm, 1984; Brenowitz and Kroodsma, 1996), how the details of the songs were learned during an early critical period (Marler, 1997), how these sounds were produced by the biomechanical details of the syrinx (Greenwalt, 1968; Podos, 1996; Fee *et al.*, 1998; Suthers, 2001), as well as understanding the fitness benefits that accrued from complex song (Searcy and Anderson, 1986).

Without information from Tinbergen's four main questions our interpretation of the biology of any behavior is incomplete. But I would like to make a further argument: without integration our interpretations are also likely to be inaccurate. A virtue of an integrative approach is that data and interpretations drawn from one level of analysis can inform data and interpretations drawn from another level. This view is antithetical to some, who maintain that there should be hard boundaries between levels of analysis so as not to confuse them (Reeve and Sherman, 1993, 2001). And there is no question that there has been confusion (Sherman, 1988; Alcock and Sherman,1994). When we ask why song birds sound as they do, we must specify whether we refer to how the syrinx generates sounds, what fitness benefits the males derive from their serenades, how they are able to acquire such a song, or from what kinds of ancestral sounds did these melodies evolve. But to pretend that these boundaries of scientific interest are anything more than human categorical constructs to assist in perceiving an otherwise incomprehensibly large subject is to risk real answers to the real question of why animals behave as they do.

In this chapter I will use studies my colleagues and I have conducted on acoustic communication in frogs to demonstrate how an integrative approach to the biology of communication behavior, which encompasses three of Tinbergen's four questions, helps us to avoid interpretations that are not only incomplete but that are inaccurate and will hopefully lead to a complete and correct understanding of why these animals communicate as they do.

## Evolution of behavior, then and now

The study of the evolution of behavior has a history that was firmly planted within what we would now call the field of phylogenetics, and which presages much of what is being done today in behavioral evolution. Influenced by the earlier work of Heinroth (1909) and Whitman

(1898), Lorenz (1941) argued that the analysis of similarities and differences in behavior could provide data to analyze phylogenetic relationships among species. This argument had been rejected because some thought behavior too flexible and inconsistent to be a reliable taxonomic variable (Atz, 1970; Aronson, 1981). But the argument has continued, and there seems to be a consensus that behavioral data can contribute importantly to phylogenetic analysis, and might even be as reliable as other data sets (DeQueiroz and Wimberger, 1993; Wimberger and DeQueiroz, 1996).

A number of ethologists were also interested in the historical patterns that gave rise to the behavior of exant species. Darwin (1871) posited that facial expression in humans were shared with other primates by descent through a common ancestor. He even suggested that in mate attraction "the same sounds are often pleasing to widely different animals, owing to the similarity of their nervous systems" (p. 91, 1872 [in 1965 reprint]). Huxley's (Huxley, 1914) hypothesis of ritualization (see also Tinbergen, 1951) might be the classical account of the historical patterns by which behavior evolves.

The study of animal behavior, however, soon was to change drastically. It underwent a revolution initiated by W.D. Hamilton's papers on the genetical evolution of social behavior (Hamilton, 1964) and G.C. Williams' (1966) book, *Adaptation and Natural Selection*. This revolution was sustained by a number of important papers, and especially by a spasm of creativity by R.L. Trivers, which we are unlikely to see again in our field for some time. In a period of six years he published five classical papers that addressed reciprocal altruism (Trivers, 1971), parental investment and sexual selection (Trivers, 1972), sex ratio evolution (Trivers and Willard, 1973), parent offspring conflict (Trivers, 1974), and haplodiploidy and insect sociality (Trivers and Hare, 1976). When E.O. Wilson codified many of the principles of sociology in 1975 a new discipline, much more closely allayed with theoretical population genetics and selection thinking, was born. Much of animal behavior was divided into studies of underlying neural mechanisms and sociobiology. Some thought that ethology was the glue that would maintain these two disparate parts in the field. Wilson, however, made a bold prediction (Wilson, 1975, p 6):

> *The conventional wisdom also speaks of ethology, which is the naturalistic study of whole patterns of animal behavior, and its companion enterprise comparative psychology, as the central unifying fields of behavioral biology. They are not; both are destined to be cannibalized by neurophysiology and sensory physiology from one end and sociobiology and behavioral ecology from the other.*

When the dust settled most studies of naturalistic animal behavior, formerly ethology and now called behavioral ecology, concentrated on studies of current adaptive significance, Tinbergen's question of survival ability. But a parallel revolution was taking place in the field of systematics (Hull, 1988). Willi Hennig (Hennig, 1950) published a work in German in 1950, in which he argued that taxonomic relationships should reflect phylogenetic ones and offered a method based on shared derived characters for uncovering such relationships. The impact of this new approach of cladistics increased substantially with the English publication of *Phylogenetic Systematics* (Hennig, 1966). The controversy that cladistics engendered in the systematics community brought it to the more general attention of evolutionary biologists and resulted in several important books that introduced phylogenetic methods to the study of ecology, evolution and behavior (Ridley, 1983; Brooks and McLennan, 1991; Harvey and Pagel, 1991; Brooks and McLennan, 2002).

The reintroduction of phylogenetics (Tinbergen's question of Evolution) to animal behavior was in a narrower context. Initially, the interaction of the new phylogenetics and behavior was not to derive behavioral characters for phylogenetic analyses, molecular characters were beginning to reign supreme in phylogenetic reconstruction, nor was it to uncover past patterns of the evolution of behavior. Phylogenetic tools were used almost exclusively to test hypotheses of adaptation. This emphasis was especially due to Felsenstein's (1985) seminal contribution on phylogenies and the comparative method. Many tests of behavioral adaptation make predictions about the associations of traits within species (e.g., testes size and mating system). Felsenstein pointed out that in such comparisons individual species are not necessarily independent data points. Taxa exhibiting similar traits might do so because they share them through descent from a common ancestor rather than independently evolving them in response to a similar selection pressure. He outlined a method, Independent Contrasts, which can be used to estimate the degree to which traits of species have diverged from a common ancestor.

Since the reintroduction of phylogenetics to animal behavior, there have been other uses of these techniques besides testing hypothesis of adaptation (reviewed (Martins, 1996; Ryan, 1996). In the remainder of this chapter, I will examine how we use a phylogenetic analysis, Tinbergen's Evolution, to complement our studies of survival value and causation of behavior. I will emphasize how including an evolutionary analysis in our studies can change interpretations of behavior based only on studies of function and causation. This is why I will emphasize that integrating Tinbergen's four questions is not only needed for a complete understanding of the biology of behavior, but it might be necessary for a correct understanding.

### SEXUAL COMMUNICATION IN TÚNGARA FROG: SEXUAL SELECTION AND SENSORY EXPLOITATION

We have taken an integrative approach to attempt a deep understanding of the sexual communication system of the túngara frog, *Physalaemus pustulosus*. These studies themselves evolved from the behavioral ecology approach to understand the selection pressures responsible for the calling behavior series in this species. Later, we added a neuroethological perspective to uncover the neural mechanisms underlying female preferences for mating calls. Finally, we used modern phylogenetic approaches to uncover patterns of signal-receiver evolution to contrast sexual selection hypotheses of good genes, runaway sexual selection and sensory exploitation, and more recently, to explore how historical contingency can influence current brain function.

A main point of this chapter is the varied uses we can make of phylogenetics to understand behavioral evolution. A subtext is that integration is necessary for a complete and correct understanding of behavioral biology. I will illustrate this by demonstrating how our interpretations of behavioral and neural data changed when we interpreted them in a phylogenetic context.

### Behavior

This research results from an ongoing collaboration with Dr. A. Stanley Rand of the Smithsonian Tropical Research Institute. We have also collaborated with several colleagues whose contributions are duly noted throughout. Reviews of this system summarizing earlier and more recent work in this system can be found in Ryan (1985) and Ryan and Rand (2003).

There are approximately 5000 species of frogs and most of them produce a species-specific advertisement call. This call is used by males to mediate male-male interactions and to make their presence known to females. Females assess these calls and decode the information that leads to assortative mating among species and selective mating within the species.

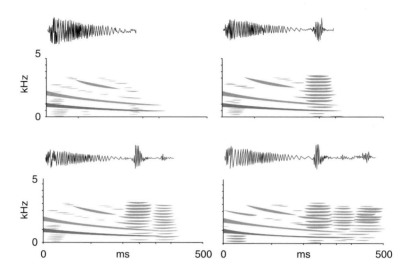

Fig. 7.1. Sonograms (bottom of each panel) and oscillograms (top of each panel) of a call complexity series of a male túngara frog. Calls contain a whine and from 0–3 chucks (proceeding clockwise from top left).

Túngara frogs, *Physalaemus pustulosus*, are small (ca. 30 mm snout-vent length), and are common throughout much of the lowland tropics in Middle America and north-eastern South America. This species is unusual in that it produces an advertisement call of facultative complexity with acoustically distinct components. The basic component of the call is referred to as a whine (Fig. 7.1). It is a frequency sweep that begins at about 900 Hz and in about 300 ms descends to approximately 400 Hz. This component is necessary and sufficient to elicit phonotaxis from females; it will also elicit calling from males, although males are permissive and will respond to a variety of sounds. All other closely related species produce whines. The second component of the túngara frog's call is the chuck. It is typically a short sound of ca. 30 ms in duration with a fundamental frequency of 250 Hz and 15 or so harmonics. Most of the call energy is in the higher harmonics.

Males usually produce a simple call, whine-only, when calling by themselves. In larger choruses most if not all of the males produce complex calls, whines followed by up to seven chucks, although most complex calls have either one or two chucks. In phonotaxis experiments females are attracted to simple calls but prefer complex calls (Rand and Ryan, 1981). This selective advantage to producing complex calls raised the question of why males do not always produce them. There should be some cost to offset the benefit. Although calling is energetically costly, adding chucks does not increase oxygen consumption or lactate concentrations (Bucher *et al.*, 1982; Ryan *et al.*, 1983). But there is a predation cost. The frog eating bat, *Trachops cirrhosus*, locates frogs by passively orienting to the advertisement call (Tuttle and Ryan, 1981). The bats respond to male túngara calls in much the same manner as do female túngara frogs. The bats are attracted to simple calls but prefer complex ones. The difference is that the bats eat rather than mate with the signaler (Ryan *et al.*, 1982).

In túngara frogs, as with a number of other species of frogs, females prefer to mate with larger males. This mate choice is mediated to a large extent by variation in the chuck of the male's call. Larger males produce chucks with a lower fundamental frequency, probably because their larynges are bigger and therefore vibrate at a lower frequency. Females prefer lower-frequency chucks and this results in them choosing larger males, and generating sexual selection for larger males, lower frequency chucks, and larger larynges (Ryan, 1980; Ryan, 1983; Wilczynski *et al.*, 1995).

Is the female's preference for lower frequency chucks adaptive? It appears to be so. In these frogs, as in most others (Shine, 1979), females are larger than males. On average, when a female chooses a larger male she reduces the size difference between her and her mate. The smaller the size difference between the two, the more eggs fertilized (Ryan 1983, 1985). Thus female choice of larger males is adaptive. This is true not only for túngara frogs but for some other species as well (Davies and Halliday, 1978; Bourne, 1993).

These studies of the behavioral aspects of the túngara frog's reproductive communication system lead to some basic conclusions. The male's call complexity series evolved under the contrasting selection pressures of sexual selection and predation–parasitism. From the female's perspective, they prefer lower frequency chucks which results in them mating with larger males. By mating with larger males they gain a reproductive advantage in having more eggs fertilized due to the smaller difference in body size between the mates. It would seem logical then that females evolved preference for lower frequency chucks as a response to selection for increased fecundity.

## Neurobiology

Together with our colleague Dr. Walt Wilczynski of the University of Texas, we decided to investigate some of the neural mechanisms that influence mate preference in females. Capranica and his colleagues (Capranica, 1977; Wilczynski and Capranica, 1984) had shown that the frog's auditory system was a valuable model for how animals decoded biologically relevant acoustic pattern, in this case the species recognition calls. We decided to extend this neuroethological paradigm for species recognition to sexual selection in order to understand how the auditory system results in certain female preferences among conspecifics.

Much of the frog's initial processing of auditory cues takes place at the periphery. There are two auditory end organs responsive to sonic frequencies. Among a suite of differences between these end organs, the most apparent is the frequency range to which each organ is maximally sensitive. The amphibian papilla (AP) is more sensitive to lower frequencies, usually below 1200 Hz, and the basilar papilla (BP) is more sensitive to higher frequencies, usually above 1200 Hz. Capranica's matched filter hypothesis (1977) suggested that the emphasized frequencies in the species' advertisement call matches the sensitivities of the two end organs. Gerhardt and Schwartz (Gerhardt and Schwartz, 2001) reviewed the literature and confirmed this hypothesis. Although many species of frogs have advertisement calls with most energy restricted to the frequency range of only one peripheral end organ, there is a strong correlation between the call's emphasized frequencies and the tuning of the end organs among species.

In túngara frogs the two call components have energy that is mostly distributed within the sensitive range of either the AP or the BP. The whine has a dominant frequency of about 700 Hz, which is close to the maximum sensitivity of the AP, and the chuck emphasizes the energy in the upper harmonics, with an average dominant frequency of about 2500 Hz, close to the sensitivity of the BP (Ryan et al., 1990). Thus whereas some species exhibit a match between the advertisement call's emphasized frequencies and tuning of one or both of the peripheral end organs, in túngara frogs each component matches the tuning of one of the end organs.

Guided by the neurophysiological data, we conducted a series of phonotaxis experiments in which we deconstructed the whine and the chuck to determine the salient features of each that were necessary to elicit phonotaxis from females (Rand et al., 1992; Wilczynski et al., 1995; Fig. 7.2). Given the distribution of energy in natural calls, only the fundamental frequency of the whine is necessary to influence phonotaxis, the upper harmonics do not influence female behavior. Of the fundamental frequency, there are three parts: one is necessary and

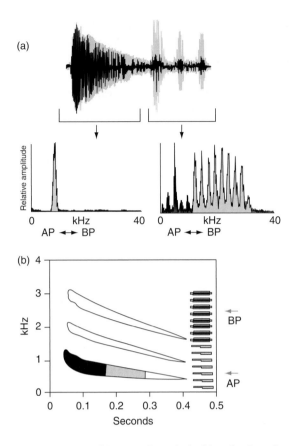

Fig. 7.2. (a) An oscillogram of a typical whine-chuck, and power spectra of the whine (left) and the chuck (right). On the x axis of each power spectrum we indicate the general range to which the AP (< 1500 Hz) and the BP (> 1500 Hz) of most anurans are maximally sensitive. (b) Illustrates a sonogram of the whine and chuck. Arrows indicate the frequencies to which the AP and the BP of the túngara frogs are most sensitive. The sonogram also illustrates the results of signal deconstruction experiments. The sounds that are not shades have no effect on female phonotaxis. Those that are shaded in black are sufficient to elicit phonotaxis, while those in gray add to the attractiveness of the signal but by themselves are not sufficient to elicit it.

sufficient to elicit phonotaxis, another section adds to the whine's attraction but by itself will not attract females, and the final part has no influence on the female. In the chuck, only the higher-half harmonics, >1500 Hz, increased the attractiveness of the chuck, and that effect could be mimicked with a pure tune near the most sensitive frequency of the BP. Combining the neurophysiological results with the behavioral experiments we concluded that in túngara frogs that AP is the end organ primarily responsible for the initial decoding of the whine, while the BP is primarily responsible for initial decoding of the chuck. These data also offer strong support, and a different kind of support, for Capranica's matched-filter hypothesis.

There is a slight mismatch between the average tuning of the BP and the chuck's dominant frequency. The average dominant frequency is about 2500 Hz while the average BP tuning is about 2200 Hz. Computer simulations confirmed the obvious expectation that lower frequency calls elicit more neural stimulation from the BP than higher frequency calls. This hypothesis is also confirmed by behavioral studies that show single tones within the sensitivity range of the BP can mimic the effects of the entire chuck, and a lower frequency tone that matches the tuning is preferred over a higher frequency tone (Ryan et al. 1990).

We can add our interpretations from the neurobiology studies to behavioral studies. Females prefer larger males because they prefer lower frequency chucks, and they prefer lower frequency chucks because these calls better match the tuning of their BP. Since the females gain a reproductive advantage from mating larger males, we assume that both the preference for complex calls, the preference for lower frequency chucks, and the tuning of the BP evolved to maximize female fecundity.

## Evolution

A comparative approach can allow us to test hypotheses about the evolution of this communication system. *Physalaemus pustulosus* is a member of the *Physalaemus pustulosus* species group. This group, as defined by Cannatella and Duellman (1984) and Cannatella *et al.* (1998), contains six species (Fig. 7.3). Three species are found in Middle America (*P. pustulosus*) or in South America east of the Andes (*P. pustulosus*, *P. petersi*, *P. freibergi*; the taxonomic status of *P. petersi* and *P. freibergi* is uncertain, and here we treat them as single species), and three are found west of the Andes (*P. coloradorum*, *P. pustulatus*, and an undescribed species, species B). It is likely that other new species will be described. The remainder of the genus, 35 or so species, are in South America east of the Andes.

As described in detail above, *P. pustulosus* produces an advertisement call of varying complexity. *P. petersi* can also make complex calls

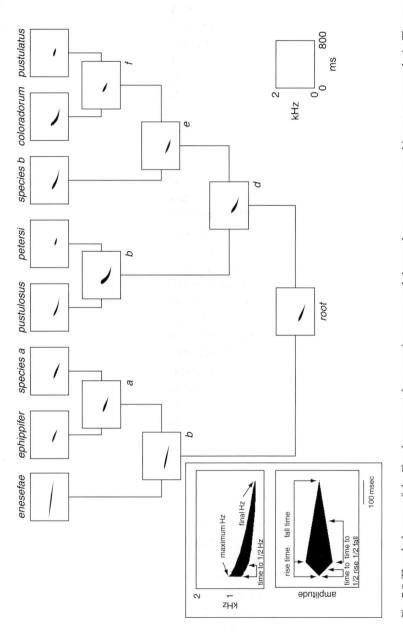

Fig. 7.3. The phylogeny of the *Physalaemus pustulosus* species group and three other congeners used in outgroup analysis. The synthetic advertisement call for each species is shown at the tips of the phylogeny. The insert shows the parameters of the fundamental frequency measured for the call of each species. These are the variables that were used for estimating and synthesizing the ancestral calls, which are shown on the nodes of the phylogeny.

(Boul and Ryan, 2004). Unlike *P. pustulosus*, these males add only one and not multiple secondary components, called a squawk, and not all populations produce complex calls. None of the species on the western side of the Andes are known to produce complex calls, and our studies of the other 30 or so other congeners not in the species group also reveal a lack of complex calls. The most parsimonious assumption is that the complex call evolved in the ancestor of the eastern clade of the *Physalaemus pustulosus* species group.

One of the conclusions about the behavior of the túngara frog's communication system was that the preference for complex calls was adaptive because preference for chucks allowed females to assess male size, and females benefited from choosing larger males (Ryan, 1983, 1985). It would seem logical then that the preference for chucks evolved in concert with the chucks. This would also be consistent with the two most popular hypotheses for the evolution of female mating preferences: runaway sexual selection, and choice for good genes. In the former, Fisher (1930) suggested that once a genetic correlation, or linkage disequilibrium, is established between genetic variation for traits and genetic variation for preferences, the prefer- ence would quickly "runaway" to fixation in a population. This is not a result of direct selection on the preference, but a correlated response of the preference to evolution of the male trait which is under direct selection (generated by female choice). The evolution of preference for good genes comes about in a similar way (Kirkpatrick, 1982). In this scenario genetic variation for the preference becomes correlated with the good genes, and again the preference runs away and becomes established in the population.

In the two scenarios described above, runaway sexual selection and good genes, the traits and preferences, or more generally, the signals and receivers coevolve. This hypothesis can be tested using phylogenetic techniques (Ryan, 1990). Let us consider a simple exam- ple in which there is a derived trait, such as a chuck, found in one taxon but absent in other close relatives. The species with the chuck also have a preference for the chuck. Since the chuck is only present in one taxon the most parsimonious assumption is that it evolved in that lineage. What about the preference? We must know something more about the preference in other species to speculate about its pattern of evolution. Not surprisingly, most studies of sexual selection have not been con- cerned about preferences for traits that do not exist. If the preference exists only in the species (or the lineage) with the chuck, we would conclude that the chuck and the preference for chuck coevolved. If instead, the preference for chuck existed in other species, even though those species were lacking the derived trait, we might conclude that the preferences of these species were shared through a common ancestor.

We tested this hypothesis of pattern of trait-preference evolution in *Physalaemus* by determining if a species in the western clade of the *Physalaemus pustulosus* species group has a preference for chucks added to their own call, even though our evidence strongly suggests that the chuck evolved in the eastern clade of the species group after these two clades diverged. We added chucks digitally excised from a *P. pustulosus* call to the call of *P. coloradorum*. When female *P. coloradorum* were given a choice between their simple call typical of their species, a whine with no chucks, versus an artificial complex call, a *P. coloradorum* whine with *P. pustulosus* chucks, females showed a preference for the latter (Ryan and Rand, 1993). This study shows that the preference for complex calls exists in species lacking such calls. The most parsimonious interpretation about the evolution of preference for chucks is that it is shared by *P. coloradorum* and *P. pustulosus* through a common ancestor. If true, then our conclusions about the pattern of evolution is that the preference for chucks existed prior to the evolution of chucks, and that males evolved chucks to exploit these preexisting preferences.

It might seem odd that a preference for such a specific acoustic trait as a chuck could exist before the chuck evolved. But it seems that the preference for the chuck is just one sample of a more general preference that includes white noise in the chuck's amplitude envelope, various pure tones, and even artificial "bells and whistles," although there certainly are sounds that do not enhance call attractiveness, and none of these other stimuli is more attractive than a chuck (Ryan and Rand unpublished data).

We can apply this phylogenetic approach to data on the tuning of the auditory periphery that is involved in decoding of the call. The most sensitive frequency of the AP matches the dominant frequency of the whine in túngara frogs. The most sensitive frequency of the BP is a close match to the dominant frequency of the chuck, and the slight mismatch between the means of both is consistent with the female túngara frog's preference for lower frequency chucks. With the exception of *P. petersi* and *P. freibergi*, other species in the *Physalameus pustulosus* species group lack a secondary component although they all produce whine-like advertisement calls. We examined the tuning of the auditory periphery of the other members of the species group to determine if the BP tuning that matches the chucks is a result of coevolution of

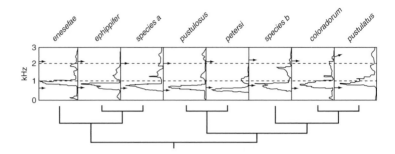

Fig. 7.4. Tuning of peripheral end organs in the *Physalameus pustulosus* species group. For each species we show the mean peak sensitivities of the amphibian papillae (lower arrow) and the basilar papillae (upper arrow). We also show a power spectrum of a typical advertisement call. For *P. pustulosus* the power spectrum is of a simple call. (Redrawn from Wilczynski *et al.*, 2001).

signal and receiver, or if the tuning is a property of the species group which males exploited when they evolved chucks.

The results strongly support the hypothesis of sensory exploitation (Ryan *et al.* 1990; Wilczynski *et al.*, 2001; Fig. 7.4). For all species the tuning of the AP matches lies with a region of substantial energy in the species' whine. AP tuning varies significantly among species, although it is not significantly correlated with the whine's dominant frequency. In the BP, however, the tuning is statistically similar among all species with one exception. *P. pustulatus* is the only species in which the BP tuning shows a significant difference from that of other species group members; its BP tuning is significantly higher. The rest of the species, including *P. pustulosus*, are not significantly different. Thus the BP tuning is a conserved property of the auditory system of these frogs. We have to reject the hypothesis that BP tuning is *P. pustulosus* coevolved with the chuck. Instead, it appears that the relationship between the chuck's spectral characters and this neural property is a result of signal and not receiver evolution. Thus sensory exploitation seems to explain both the behavioral preference for the chuck as well as some of the tuning properties of the peripheral auditory system.

## Summary

In our studies of sexual communication in túngara frogs we have attempted to understand the role of sexual selection in the evolution of the complex advertisement call. We have shown that behavioral studies have documented a selective advantage as well as the selective trade-offs for males making complex calls. The studies of neurobiology identify some of the underlying mechanisms that enable females to decode the whine and the chuck and also suggest a mechanism to explain the female's behavioral bias toward lower frequency chucks. Initially, the evolutionary interpretation was that females evolved a preference for complex calls and low-frequency chucks because it allows them to choose larger males and thus maximize reproductive success. Thus trait and preference, signal and receiver, must have coevolved to bring about such benefits. The addition of an evolutionary perspective, however, changes this interpretation. Both the preference for chucks and the auditory tuning that guides females to lower frequency chucks are not restricted to species with chucks; both of these receiver traits are found in species that produce only simple calls. Thus it appears that there was a preexisting preference for chucks and that in evolving chucks males exploited this pre-exisiting preference.

Since we first suggested the hypothesis of sensory exploitation (Ryan, 1990; Ryan et al., 1990) a number of other studies have suggested similar results, including studies of swordtails; auklets; spiders, water mites, song birds, (summarized in Ryan, 1998). These findings impact our understanding of sexual selection, but are applicable to the more general field of animal communication (e.g., Kilner and Davies, 1999), and are readily applicable to other fields such as cognitive psychology (Enquist and Arak, 1998) and conservation biology (Schlaepfer et al., 2002).

SEXUAL COMMUNICATION IN TÚNGARA FROG:
SPECIES RECOGNITION AND HISTORICAL
CONTINGENCY

Up to now we have been addressing how female's use variation in male signals within the species to choose mates, and the evolutionary pattern by which the current relationships between signal and receiver came into being. Females also use advertisement calls to ensure conspecific matings. Most frogs call, and all known advertisement calls differ among species. Furthermore, in phonotaxis experiments females almost always show a preference for the conspecific call over a heterospecific call, even if the females are allopatric with heterospecific. There are innumerable strategies a receiver, human, frog, or computer, can use to discriminate between pairs of stimuli. This is because most animal signals are composites of suites of characters and different weighting schemes (i.e. the value of individual parameters in decision making) can produce the same results; the study of feature weighting is an important part of animal communication (Nelson and Marler, 1990; Ryan and Getz, 2000). This has been shown in a study by Gerhardt (1981) of two species a tree frog: *Hyla gratiosa* and *H. cinerea*. These two species have similar advertisement calls, and they are both sympatric and synchronic with one another. For each species, the most challenging reproductive decision is decoding the differences between this pair of sounds. By painstakingly varying individual signal parameters, Gerhardt showed that each species weighted call parameters differently when making the same discrimination. In terms of natural selection and adaptation this difference in recognition strategies might seem trivial; the end result is that females prefer conspecifics over heterospecifics. But clearly, the way in which the two species achieve the same function differs; their brains work differently and we would like to know why.

We have recently been investigating the evolution of species recognition in túngara frogs. Again we have utilized an integrative approach in which studies of behavior and neurobiology (and computational neurobiology) are performed in a phylogenetic context. The specific question is how does past history influence species recognition decisions, the more general question is how does historical contingency influence brain function.

## Behavior

The túngara frog is allopatric with all member of the *Physalaemus pustulosus* species group, and with all members of the genus with one exception: túngara frogs and *P. enesefae* overlap in a small region of the llanos in Venezuela. Nevertheless, the expectation is that female túngara frogs should prefer their advertisement call over that of other species.

We conducted a series of phonotaxis experiments in which we synthesized the average call of each species of the species group as well as those of three congenerics not in the species group. We asked if females would discriminate between the two calls, showing a preference for the conspecific call. These calls contained only the fundamental frequency of the whine since in túngara frogs information in the upper harmonics of the call do not influence female phonotaxis. In all cases female túngara frogs showed a statistically strong preference for the conspecific call.

We also asked if females would recognize the calls of the other species by presenting them the conspecific call paired with a white noise stimulus. This answer can not be garnered from the discrimination experiments. For example, we know that female túngara frogs prefer a whine-chuck to a whine-only. These results do not reveal whether the whine by itself is recognized by females as an advertisement call.

When we conducted the recognition tests we found that females showed statistically significant recognition for a number of heterospecific signals. Why? One reason is that there has been no selection to avoid such calls since túngara frogs are allopatric with these heterospecifics. In the absence of selection, we might expect that calls that sound similar to that of the túngara frog might elicit a response from female túngara frogs. Thus overall call similarity might explain the cases of false recognition.

Another possible explanation is that of historical contingency. Túngara frogs share recent ancestors with the heterospecifics we tested. Common ancestry explains why species in the group share similar call characters, such as frequency modulated whines. Shared ancestry could also suggest that females of different species share some similar auditory biases. If females of one species find a certain call appealing it should not be surprising that closely related females might as well, especially when the species are allopatric and there has been no selection for avoiding these heterospecifics (Ryan *et al.*, 2003). We thought we could disentangle the roles of overall acoustic similarity. Our phylogeny of the *Physalaemus pustulosus* species group was derived from several data sets: gene sequences, allozyme variation, morphological characters and advertisement calls. Analysis of each data set yields a similar set of phylogenetic relationships with one exception: calls. A phylogeny based on call variation alone little resembles the consensus tree or any of the other phylogenetic hypotheses (Cannatella *et al.* 1998).

We used various techniques to reconstruct ancestral calls for the species group. For each call variable we estimated its value at the ancestral nodes of the tree using Felsenstein's Independent Contrast method (Felsenstein, 1985), as well as other methods of estimation. We then used the estimates of each of the call variables at each node to synthesize the ancestral call. This exercise, we realize, does not guarantee a good estimate of what ancestors sounded like. We do feel, however, that these estimates give us some idea of the acoustic landscape traversed by ancestors in the species group during the evolution of call recognition. A major caveat is that since calls are not highly phylogenetically informative such an exercise might be doomed from the start, but as we will see when we present the study of the artificial neural networks, we can show that out estimates of ancestors are significantly better than random.

We repeated the discrimination and recognition tests with the ancestral calls, and a pattern began to emerge. Although females mostly still discriminate in favor of the conspecific call, the strength of the discrimination shows some variation among calls. In particular, females do not show statistically significant discrimination between the conspecific call and the call at the ancestral node with *P. petersi*. In the recognition experiments there are a large number of calls, ancestral and heterospecifics alike, that females falsely recognize as indicating an appropriate mate.

Due to overall acoustic similarity and phylogeny not being correlated we can partition the effects of each in explaining how female responses vary among calls. In the discrimination experiments phylogenetic relatedness explains 45% of the variation in female phonotactic responses while call similarity only explains 18%. In the recognition experiments both phylogeny (38%) and call similarity (31%) contribute similarly to explaining why females do what they do (Ryan and Rand, 1995, 1999).

We interpret these results as evidence that historical contingency influences the strategies that female túngara frogs use to decode species specific calls, that is, to decide if a call indicates an appropriate mate. This is probably because ancestors of túngara frogs shared not only similar calls but similarities in how the brain decoded these calls. This interpretation also generates a hypothesis, although one that did not seem testable at the time: frog brains that evolve to decode different calls in their past will use different decoding strategies to decode the same call in the future. In short, historical contingency influences brain function.

## Artificial neural networks

The hypothesis about brain evolution influencing decoding schemes seemed one doomed to be evaluated by the strength of its logic rather than experimental testing. Artificial neural networks, however, offer a way to test this hypothesis.

In a previous study we had trained populations of artificial neural networks to recognize túngara frog calls (Phelps and Ryan, 1998, 2000; Phelps *et al.*, 2001). We showed that the response of the networks to novel stimuli predicted the response of female túngara frogs to the same novel stimuli, suggesting that there are some similarities in which the two systems decode the target stimulus and generalize to other stimuli. We used this approach to test the hypothesis of historical contingency described above.

Three populations of artificial neural networks were trained to recognize a series of three calls before finally being trained to recognize the túngara frog call. In one population, those three calls were the ancestral calls from the root of the species group phylogeny and the two direct descendents of the túngara. The nets in the populations were trained to recognize the root call. Once the fitness criterion was reached, they were trained to recognize the immediate descendent of the root. Once the net achieved recognition of that call, they were trained to recognize the immediate descendent of the node ancestral to *P. pustulosus- P. petersi*. After recognition of this call was achieved, the nets were trained to recognize the túngara frog call. We refer to this population as having a mimetic history. A second population of nets was trained to a random history. Three calls were chosen at random from the species group, including both ancestral and heterospecific calls. After being trained to recognize each of the three calls in sequence the nets were then trained to recognize the túngara frog call. A third population has a mirror history. Here, the three ancestral calls used in the mimetic history were rotated 180 degrees in multivariate space creating "mirror images" of the calls. Although to the human ear these calls did not sound very different from the other ancestral calls, they occurred in a portion of the acoustic landscape not utilized by the species group, heterospecifics or ancestors. We used this history because the calls used in training had the same degree of acoustic differences as in the mimetic history, but the calls themselves were different.

The first result of these neural network simulations showed that the networks' past history did not constrain its ability to evolve to recognize the túngara frog call. We then asked if the networks influenced the manner in which the nets decoded the call. We did this by determining the degree to which each population of networks could predict the behavior of females. As in our earlier studies the response of the trained networks and the female túngara frogs to a variety of novel stimuli was determined. Only the nets with the mimetic history showed a significant correlation between their response and the response of real females. A maximum likelihood analysis also showed that the response of the mimetic history nets was a significantly better predictor of female preference than was either the nets with the random history or the mirror history.

The neural network simulations show several results. One is that history does not constrain the evolution of recognition strategies. A second is that history influences the decoding strategy used. Third is that the neural networks with a mimetic history must have utilized a computational strategy in some ways similar to the one used by female túngara frogs.

We hope that we have not been seduced by the elegance of artificial neural networks; we realize artificial intelligence machines are not animal brains. We feel, however, that we offer an approach to testing hypotheses that can be added to an arsenal of other techniques. We think they are especially useful for investigating historical contingency because they capture some of the complexity and the unpredictability as to how brains come about to solve problems.

### SUMMARY AND CONCLUSIONS

Tinbergen (1963) outlined an important heuristic for studying the biology of behavior. His four questions categorized different approaches that can be taken to understand behavior, and it is obvious that all of these questions must be addressed for a complete understanding of behavior. But there is more to it. I argue that all of these questions must be addressed for a correct understanding of behavior, and I illustrate this point with our studies of sexual communication in túngara frogs. Our studies of sexual selection and neural mechanisms, more generally, survival ability and causation, led to a logically consistent interpretation of how this communication system evolved: trait and preference coevolve to maximize reproductive success. It was logical but was it true? The addition of a phylogenetic component to our studies showed it was not. Trait and preference did not coevolve but instead traits evolved to exploit pre-existing preferences. Furthermore, our studies showing female recognition of ancestral and heterospecific calls could easily have been interpreted as females generalizing to calls that were similar in their overall acoustic structure, but again by including Tinbergen's question of evolution we see that the vagaries of past history influence the way the frog's brain works today.

My task in this chapter was twofold. First, to illustrate some of the ways ethologists are currently addressing the question of evolution. Second, to argue that employing the four aims and methods of Tinbergen is not sufficient, but these aims and methods need to be integrated to have an understanding of the biology of behavior that is both correct and complete.

# 8

# Do ideas about function help in the study of causation?

In his 1963 paper on the aims and methods of ethology, Tinbergen credits Julian Huxley for identifying *causation*, *survival value*, and *evolution* as the three major problems in biology. To this list of three, Tinbergen added the problem of ontogeny. The distinction that Tinbergen described among causal, functional, developmental, and evolutionary questions is one of the most lasting legacies of ethology. The idea that each of these questions can be asked of any behavior, and that each requires its own distinctive answer, continues to shape the way we think about behavior. Tinbergen did not believe these questions stood in isolation and never interacted. He pointed out, for example, that we can ask about the cause, function and evolution of development (Tinbergen, 1963). Contemporary behavioral ecologists often make the point that the cause of behavior is easier to analyze and understand and if its function is known (Krebs and Davies, 1997; Stephens and Krebs, 1986). This latter idea has been criticized, however, for confusing the essential difference between causal and functional explanations of behavior (Bolhuis, 2005, this volume; Bolhuis and Macphail, 2001; Macphail and Bolhuis, 2001).

The reasons for drawing a distinction between causal and functional explanations of behavior are so clear and so familiar that it is surprising the two are sometimes confused by students and, occasionally, by professional researchers. The difference between causal and functional explanations can be illustrated with a non-behavioral example. The function of a hammer is to drive nails. Indeed, it is function that makes an object a hammer, to use one of philosopher Martin Heidegger's favorite illustrations (Heidegger, 1996). The causes of a hammer have nothing to do with driving nails but instead involve shaping a piece of wood to make a handle, forming stone or metal into a hammer head, and fastening the two pieces firmly together. No matter

*Tinbergen's Legacy: Function and Mechanism in Behavioral Biology*, Johan J. Bolhuis and Simon Verhulst (eds.). Published by Cambridge University Press. © Cambridge University Press 2008.

how important the function of a hammer may be, it cannot cause a hammer to come into being. In addition, a function performed in the future, like driving nails, cannot act at an earlier point in time to bring a hammer into existence. Nevertheless, we have all read student essays and even papers in the scientific literature with statements like, "Better over-winter survival in the tropics causes migratory birds to leave the breeding grounds early in autumn." Why does this confusion exist?

There are at least two reasons that otherwise sensible people confuse causal and functional explanations of animal behavior. The first is that we, as humans, can consciously anticipate the future function of our behavior and planning for the future can be a cause of behavior. Because I can foresee needing a hammer tomorrow I can make one today, or at least go out and buy one. This is not a case of a function propagating backward in time to cause behavior but instead the outcome of learning and past experience acting as a cause. As humans we are used to foreseeing functions and acting accordingly and this way of thinking is sometimes applied – surely without giving the matter much thought – to animal behavior. As far as we know, animals cannot do this. There may be some instances, such as chimps and other primates carrying rocks to places where they will later use them to break open nuts (Boesch and Boesch, 1984; Cleveland *et al.*, 2004; Mercader *et al.*, 2002) and it is possible, though not proven (Jalles-Filho *et al.*, 2001), that there is something shared here between human and non-human primate cognition. For the most part, however, we have little evidence that anticipation of future function serves as a cause of behavior in animals.

The second reason for confusion about the cause and function of animal behavior is the theory of natural selection. A statement such as, "Selection for over-winter survival causes birds to migrate early in autumn" is an example. Sometimes this is a genuine confusion. Sometimes it is sloppy writing (or sloppy thinking). But it is not always complete illogic. Sometimes it is simply shorthand for, "Natural selection, in the form of differential over-winter survival, acts on heritable variation in the timing of autumn migration. The result is greater frequency in the population of those causal mechanisms that lead to early autumn migration." Such shorthand is widely used and understood in behavioral ecology (Cuthill, 2005).

Tinbergen was, of course, aware of the fallacy of mistaking functional explanations for causal explanations and was also aware of some more subtle pitfalls for the unwary concerning cause and function (Tinbergen, 1963; see also other chapters in this volume). We identify behavior for scientific investigation by its function. We study mate choice, foraging, migration, flight, parental care, anti-predator behavior, communication and similar functional categories. Research that specifically addresses broad causal mechanisms, such as learning, often does so in a narrower context that is functionally defined, such as avoidance learning, spatial learning, imprinting, or song learning. The behavior that seizes our interest is usually not behavior that is defined descriptively, topographically, or because it shares causes with other behavior, but instead behavior that serves some readily identifiable function in the life of the animal. Tinbergen (1963) felt that one of the greatest contributions of ethology to the life sciences was:

> "... to consider behaviour patterns (and by implication the mechanisms underlying them) as organs, attributes with special functions to which they were intricately adapted. This again facilitated causal analysis without interference by subjectivism or teleology (p. 413)"

> "The treatment of behaviour patterns as organs has not merely removed obstacles to analysis, it also positively facilitated causal analysis ... (p. 414)"

There are hazards in this functional approach though, and one hazard that Tinbergen pointed out is that although we usually identify and label behavior by its function it is not true that behavior serving the same function in different species necessarily shares the same causes (Tinbergen, 1963). This may seem an obvious point, but there is an ever-present temptation to suppose that behavior we identify as "communication," "parental care" or "food hoarding" in different animals has common neural, hormonal or energetic causes in these animals. This need not be the case, even in closely related animals. Recognition of mating calls in the treefrogs *Hyla cinerea* and *H. gratiosa* for example, is based on quite different auditory features (Gerhardt, 1981, 2001). Female preferences and selectivity in their response to male calls can be investigated experimentally by observing the phonotactic responses of females to the playback of recorded male calls and synthetic calls. Although the mating calls of males of both species have an abrupt pulsed onset, female *H. cinerea* do not discriminate between calls that have this pulse and calls that do not, while female *H. gratiosa* do. *H. gratiosa* females are selective for the low frequency component of a call while *H. cinerea* females are not. Female *H. cinerea* discriminate against male calls with added mid-frequency components while female *H. gratiosa* do not (Gerhardt, 2001). The function of the male call is identical in both species and the treefrogs share all but their most recent evolutionary history, but the acoustic features used for call recognition are quite different.

A second more subtle hazard noted by Tinbergen is that behavior we identify as a discrete functional unit may be not be organized as a discrete unit causally. As observers we see smoothly integrated functional units of behavior but that does not mean that the nervous system organizes things in that way. Recognizing an object and reaching out to grasp it seems a simple integrated unit of behavior but is, in humans at least, brought about by two separate visual processing streams, one serving perception and another serving the reaching and grasping action (Goodale and Westwood, 2004).

I will argue that on balance, understanding the function of behavior, or even having a plausible hypothesis about the function of behavior, facilitates causal analysis. It can do this in at least three different ways. As described above, function defines the categories of behavior about which we ask causal questions. In addition, discoveries about the function of behavior raise novel questions for causal analysis. These are the "clues" about causation described by Bolhuis (2005). Finally, functional considerations set the criteria that causal explanations of behavior must satisfy.

Is it possible to synthesize the study of cause and function? In reading some of the historical literature in ethology to prepare for this symposium I could find only one ethologist who was rash enough to suppose it might be possible to synthesize the study of cause and function into one seamless enterprise, but this radical ethologist was unique in many ways; more about him later.

NOVEL CAUSAL QUESTIONS

### The major histocompatibility complex

Discoveries about function can raise new questions for causal investigation. The idea that kinship and relatedness play a significant role in animal social behavior revolutionized the study of behavior from the 1960s onward (Hamilton, 1964; Trivers, 1972). Inclusive fitness provided a powerful new theoretical framework for the study of animal social organization and directed attention to behavior such as cooperative care of the young that was previously noted primarily for its novelty rather than its theoretical importance. There was also a revival of interest in inter-sexual or "epigamic" sexual selection, mate choice, and the possibility that animals may choose mates on the basis of "good genes" (Andersson, 1994; Neff and Pitcher, 2005). But these phenomena immediately raise a host of causal questions. How do animals recognize kin?

How do animals determine whether a potential mate possesses genes that will increase offspring reproductive success? There are many ways, both simple and complex, that animals can distinguish kin from non-kin (Holmes, 2004; Holmes and Sherman, 1983; Sherman *et al.*, 1997) and assess the suitability of a mate. One of the most unexpected causal mechanisms, however, is the discovery that animals can discriminate by odor among individuals differing only at loci of the major histocompatibility complex (Penn and Potts, 1999; Singer *et al.*, 1997).

The major histocompatibility complex in mammals is a set of about fifty highly variable linked genetic loci that play a central role in the immune response and self/non-self recognition. MHC Class I genes code for cell surface proteins in most cells while MHC Class II genes code for cell surface proteins of antigen-presenting cells such as macrophages. The proteins coded for by these genes bind to peptide antigens which are then presented at the cell surface for immune system surveillance. Helper T-cells recognize the MHC plus antigen complex and initiate an immune response to the antigen. Most Class I and Class II MHC loci, called the H-2 loci in mice and the HLA loci in humans, are highly polymorphic with hundreds of alleles at some loci. The particular combination of alleles occupying the linked MHC loci on a chromosome is called the MHC haplotype. The extreme allelic diversity that occurs at the MHC loci produces differences in MHC haplotype among all but the most closely related individuals (and is the reason that unrelated individuals are rarely suitable as organ donors).

Research by Yamazaki and his colleagues (Yamazaki *et al.*, 1976) first showed that inbred strains of mice have disassortative mating preferences for MHC haplotype. Mice of some strains, but not all, prefer to mate with individuals differing from themselves at MHC loci. In the initial experiments of Yamazaki *et al.* (1976) male mice were caged with pairs of estrous females that were genetically identical to the males except at MHC loci. In these tests, one female shared all or part of her MHC haplotype with the male and one female did not. Males mated preferentially with the female that differed from themselves in MHC haplotype. Subsequent research has shown that female mice, too, discriminate among potential male mates that differ only at MHC loci. When offered a choice, females prefer to mate with MHC-dissimilar males. Cross-fostering studies show that MHC preferences are learned during a period of association with parents and siblings in the nest (Penn and Potts, 1998). Mice prefer as mates individuals that are MHC-dissimilar to the animals with which they were raised, and that can mean, for experimentally cross-fostered mice, that they prefer as mates individuals of their own MHC haplotype.

Disassortative mating preference for MHC genotype has a number of consequences for mice, including avoidance of inbreeding, and promotion of pathogen resistance in offspring both through MHC heterozygosity itself and by creating a moving target for pathogens by maintaining very high levels of MHC variation (Penn, 2002).

In addition to disassortative mating preferences, there are circumstances in which animals prefer to associate with individuals that are similar to themselves at MHC loci. Female mice, which will nest and nurse pups communally, prefer to nest with other females of similar MHC haplotype (Manning *et al.*, 1992) and there is mutual attraction between MHC-similar mothers and pups (Yamazaki *et al.*, 2000). Attraction to individuals of the same MHC genotype in communal nesting and parental care by female mice is a form of kin recognition (Penn, 2002).

Mice detect and respond to the MHC identity of other mice using odor cues that are affected by the MHC loci (Singer *et al.*, 1997). Mice can discriminate by odor inbred mice that differ only at the MHC loci and can also discriminate the volatile components of the urine of these individuals. Chromatographic analyses show that it is not unique volatile components of urine that distinguish MHC haplotypes but instead the relative proportions of the many volatile components of urine (Singer *et al.*, 1997). How MHC genotype influences the odor of mouse urine is not entirely clear. MHC molecules may bind protein fragments of odorous molecules and in this way influence their relative proportions in urine. It is has been suggested that gut microflora are the source of some of the molecules bound by MHC molecules.

All of the research on the MHC loci described so far concerns inbred strains of mice, selected to be genetically uniform at all loci except the MHC loci of interest. This confers experimental control over MHC similarity, but at the cost of realism. It is therefore reassuring that discrimination among individuals and mating preferences based on MHC loci have also been described in other animals, such as wild caught three-spined sticklebacks *Gasterosteus aculeatus* (Milinski *et al.*, 2005).

In research on the MHC loci, functional ideas about mating preferences and kin recognition raised new questions about causal mechanisms. Functional considerations gave a context and significance to the discovery that mice of the same MHC haplotype preferred not to mate. The discovery that mice and other animals can discriminate among even very closely related individuals on the basis of MHC loci provided an answer to a causal question that had been identified in the course of a functional examination of behavior.

## Memory in food-storing birds

Most species in the passerine families Corvidae (the crows and jays), Sittidae (the nuthatches) and Paridae (the chickadees and tits) store food. They place between hundreds and many thousands of food items in widely-dispersed cache sites and later retrieve these caches by remembering where they hid them (Kamil and Balda, 1985; Sherry et al., 1981; Shettleworth and Krebs, 1982; Vander Wall, 1982). The discovery that these birds can remember many scattered spatial locations, in some cases for many months, has led to a great deal of research on the properties of memory for cache sites and on how memory is implemented neurally in food-storing birds (Shettleworth, 2003; Smulders and DeVoogd, 2000).

Memory for cache sites, like mate choice and kin recognition, is a functionally defined category of behavior. The discovery that animals can perform the feat of remembering large numbers of dispersed cache sites in the wild raises a host of new causal questions about behavior. How is memory for cache sites organized, and how are cache site memories encoded, stored, and retrieved? What is remembered about a cache site, how is its spatial location specified, how is it represented neurally, and what parts of the brain are involved? Recent research on three causal aspects of memory for cache sites shows how a novel function, namely memory for cache sites, has led to novel lines of causal inquiry.

An early discovery in work with food-storing birds was that the hippocampus plays an essential role in accurate memory for food caches and in memory for other kinds of spatial information. Removal of the hippocampus severely impairs the birds' ability to find caches and perform other kinds of spatial tasks (Hampton and Shettleworth, 1996a; Sherry and Vaccarino, 1989). Cache retrieval is hippocampus-dependent. In addition, it was discovered that the hippocampus is considerably larger in food-storing birds than in non-storing species (Krebs et al., 1989; Sherry et al., 1989). These initial findings have stimulated extensive research on the avian hippocampus and its relation to memory, including descriptions of the distribution of neurotransmitters and neuropeptides in the avian hippocampus (Erichsen et al., 1991; Krebs et al., 1991), studies of seasonal change in the size of the hippocampus (Krebs et al., 1995; MacDougall-Shackleton et al., 2003; Shettleworth et al., 1995; Smulders et al., 1995; Smulders et al., 2000), and research on seasonal patterns in hippocampal neurogenesis (Barnea and Nottebohm, 1994, 1996; Hoshooley and Sherry, 2004). Long-term potentiation, the effect of stress hormones, hippocampal development, sex differences, and species differences in the hippocampus have all been examined in an effort to understand how memory is supported by the avian hippocampus (Smulders and DeVoogd, 2000).

### Hippocampal size

The first descriptions of species differences in hippocampal size showed that food-storing birds have a larger hippocampus, relative to the rest of the forebrain, than non-food-storing species (Krebs *et al.*, 1989; Sherry *et al.*, 1989). These studies compared members of the corvid, sittid, and parid families to a variety of non-food-storing birds in other passerine families. Over a wide range of forebrain size, food-storing birds had a consistently larger hippocampus.

Not all members of the corvid and parid families store food, and even among those that do, there is variation in how much storing occurs. A number of studies have examined the relation between size of the hippocampus and food storing behavior within the corvid and parid families and found that relative hippocampal size is greater in species that store more (Hampton and Sherry, 1992; Hampton *et al.*, 1995; Healy *et al.*, 1994; Healy and Krebs, 1992). Brodin and Lundborg (2003) performed a meta-analysis of these data along with new data of their own and concluded, however, that there was no relation between the level of food storing and hippocampal size among species within either the corvid family or the parid family. This finding raises a number of questions about the actual relation between food storing and the hippocampus. It is possible, of course, that the large differences between food-storing and non-food-storing passerine families described by Krebs *et al.* (1989) and Sherry *et al.* (1989) simply do not occur within food-storing families. All members of the corvid family are more closely related to each other than they are to members of other passerine families in which food storing does not occur, and the same is true for all members of the parid family. Perhaps there has not been the kind of divergence within food-storing families that is seen between food-storing and non-food-storing passerine families. It is also possible that the assignment of corvid and parid species to categories of food-storing intensity in the studies re-analyzed by Brodin and Lundborg (2003) was not accurate. A further re-analysis, however, has produced a remarkable and unexpected finding (Lucas *et al.*, 2004).

Lucas *et al.* (2004) noticed what they called an obvious but inexplicable trend in the available data: Hippocampal size is consistently greater in Eurasian corvids and parids than in North American members of these families. When "continent" is included as a factor in the statistical analysis, there is a significant continent effect, and the hippocampus is significantly larger in species that store more food both within corvids and within parids.

It is indeed unexpected that Eurasian species of food-storing birds have larger hippocampuses than North American species. The effect of this, however, is that the relation between hippocampal size and food-storing behavior that can be seen in North American birds and Eurasian birds in both the corvid and the parid families is masked when birds of different continents are pooled as Brodin and Lundborg (2003) did. Lucas *et al.* (2004) tested for possible confounding variables that might produce this continent effect, such as the time birds were held in captivity before sacrifice or differences among laboratories in processing brains and estimating hippocampal size and found little evidence that these factors were the causes of the continent effect. Instead they propose that ecological differences between Eurasia and North America, and phylogenetic differences between Eurasian and North American food-storing birds, may account for the continent effect. In North America, for example, there are no non-food-storing parids, while in Eurasia there are two. Differences between North America and Eurasia in the history of human impact on ecosystems and species diversity may have had effects on the food-storing species present in the two continents and their degree of reliance on stored food. In addition, the "continent" effect matches quite closely with a phylogenetic difference between North American and Eurasian corvids. This is not the case for parids. It is possible that this phylogenetic difference is responsible for differences in the relation between food-storing and hippocampal size between Eurasian and North American corvids. Whatever the eventual explanation for the continent effect, it is clear that in Brodin and Lundborg's (2003) meta-analysis it masked the relationship between food storing behavior and hippocampal size described in previous studies that made comparisons among birds from the same continent (Basil *et al.*, 1996; Hampton and Sherry, 1992; Hampton *et al.*, 1995; Healy *et al.*, 1994; Healy and Krebs, 1992).

### Hippocampal neurogenesis

Until very recently the prevailing view was that neurogenesis was rare or non-existent in the adult brain. Although evidence for adult neurogenesis in rats had been available as early as the 1960s (Altman and Das, 1965), it was research on neurogenesis in the adult songbird brain that stimulated the current intense interest in adult neurogenesis (Goldman and Nottebohm, 1983). New neurons are added not only to some song control nuclei but are found throughout the forebrain, including the hippocampus, of adult songbirds. In the food-storing black-capped chickadee, there is a seasonal pattern of recruitment of new neurons into the hippocampus that peaks in fall at about the time the annual food-storing cycle is beginning (Barnea and Nottebohm, 1994, 1996). Recruitment in this context refers to the presence of neurons labeled with a cell-division marker injected 6 weeks previously. The occurrence of a cell division marker, such as tritiated thymidine or bromodeoxyuridine (BrdU), in the nuclei of these cells shows that they were dividing when the marker was injected, have differentiated and migrated from the ventricular zone where they originated, and have survived for six weeks.

The seasonal pattern in recruitment described by Barnea and Nottebohm (1994) could occur for a number of reasons. One possibility is that more cells are born in early fall than at other times of year and, allowing six weeks for migration and differentiation, there are therefore more new neurons in the hippocampus in fall. Another possibility is that cells are born at the same rate year round but differential survival in different seasons leads to the fall peak in recruitment. Hoshooley and Sherry (2004) showed that the rate of cell birth did not change with season, making it more likely that differential attrition is responsible for the seasonal pattern in recruitment. This finding answers one question but raises many more. Given that the production of cells is relatively constant year round, what seasonally varying process causes the observed seasonal variation in recruitment? Recruitment and attrition could be under photoperiodic control, leading to the seasonal profile observed by Barnea and Nottebohm (1994). Exposure of black-capped chickadees to the short day length experienced in fall does not, however, lead to greater hippocampal neuronal recruitment (Hoshooley et al., 2005). It is also possible that use and disuse influence the number of new neurons that are incorporated into hippocampal circuits and that the seasonal onset of food storing directly influences hippocampal activity and hence neuronal recruitment. Hoshooley et al. (2007) found a peak in hippocampal neuronal recruitment in chickadees in January, later than that found in previous studies. This may indicate year-to-year variation, or population differences, in the timing of incorporation of new neurons into the hippocampus, influenced perhaps by birds' food-storing responses to environmental conditions. The behavior of food-storing birds may thus provide clues to the control of adult neurogenesis that not only answer questions about the natural history of food storing but also have implications for the therapeutic use of neuronal stem cells in humans (Nottebohm, 2002).

## CRITERIA FOR CAUSAL EXPLANATIONS

When the function of behavior is known, and even when there is a plausible hypothesis about function, the details of any proposed causal hypothesis are constrained. Causal explanations must meet design criteria that are set by the function of behavior.

### Auditory localization in barn owls

Barns owls can capture prey in complete darkness using only sound to determine the prey's location. The faint sounds produced by a mouse walking over the ground are sufficient for a barn owl to establish during flight where the mouse is, capture it, and fly off with it in its talons. To understand how barn owls are able to do this Konishi and his colleagues established first what functional problems the owl auditory system had to solve, then searched in the barn owl auditory system for circuits that could extract the information required to establish the spatial co-ordinates of a rustling mouse (Konishi, 1995).

A point sound source that is not perfectly aligned with the front of a listener's head – whether barn owl or human – will be detected slightly differently by the two ears. As the sound source is displaced laterally with respect to the sagittal plane through the listener, the sound will reach one ear before it reaches the other and the sound waves will arrive at the two ears out of phase. The degree to which they are out of phase provides a cue to whether the displacement is to the left or right and by how much. As the sound source is displaced vertically with respect to the horizontal plane through the listener's ears, differences in the structure of the external ear result in intensity difference that provide cues to elevation. In the barn owl, the external ears are asymmetric and modified to increase this intensity difference. The right ear is located slightly below the left ear and directed upward while the left ear is directed downward. Sounds originating above the horizontal plane are loudest in the right ear, and sounds below the horizontal plane are loudest in the left ear.

By placing miniature microphones in the ears of barn owls researchers were able to verify these asymmetries in sound reception

(Moiseff and Konishi, 1981). Barn owls will turn their head to orient to a sound source and by placing miniature speakers in the ears of barn owls, researchers could make the owls orient to phantom sound sources by varying the time and intensity differences between sounds presented to the two ears (Moiseff and Konishi, 1981).

This research has led to a detailed characterization of the auditory processing streams involved in sound localization in barn owls. Phase-sensitive auditory neurons extract information about the arrival time of sound at the cochlea of each ear. Neurons in the nucleus magnocellularis encode the phase of the sound signal from each ear. Axonal "delay lines" projecting from the nucleus magnocellularis bring the phase of the two signals into register in the nucleus laminaris where "coincidence detector" neurons fire maximally only when signals from the two ears are in phase. The subset of coincidence detectors that fires corresponds to the phase difference between the two ears and hence encodes the lateral displacement of the sound with respect to the head (Carr and Konishi, 1990; Moiseff and Konishi, 1983; Peña et al., 2001).

Intensity differences between the two ears are handled by a different auditory processing stream. The intensity of the auditory signal at each ear is relayed by neurons in the cochlear nuclei to the nucleus ventralis lemnisci lateralis pars posterior (VLVp) in the pons. Neurons in this nucleus receive an excitatory signal corresponding to loudness in the contralateral ear and an inhibitory signal corresponding to loudness in the ipsilateral ear. VLVp neurons are selective for particular loudness differences and thus code for displacement of the sound source above or below the horizontal plane. Finally, timing differences and intensity difference signals are combined in the inferior colliculus to yield a representation of the horizontal and vertical location of the sound source (Konishi, 1993, 1995).

Konishi (1995) is particularly clear that functional ideas about sound localization by barn owls guided the search for causal mechanisms. "… had the researchers not known the perceptual problems the animals must solve, they would not have looked for neurons selective for these natural stimuli (p. 270)."

### Orientation to the magnetic field of the Earth

Birds, mammals, reptiles, amphibians, fish, and a variety of insects can derive orientation cues from the Earth's magnetic field. Hatchling sea turtles, for example, maintain a bearing with respect to the earth's magnetic field to leave their natal beach and head out to sea. Migratory birds and homing pigeons orient with respect to the magnetic field, and calibrate their magnetic bearing against a bearing derived from sun compass information. Surprisingly, it is not the polarity of the Earth's magnetic field that animals use to maintain a bearing but the inclination of the magnetic field.

The magnetic field of the Earth has a north pole, a south pole, and a field of magnetic force that connects the two. At the poles the magnetic field lines are oriented perpendicular to the surface of the earth while at the magnetic equator these field lines are parallel to the Earth's surface. Between the equator and the pole, magnetic field lines have intermediate inclinations to the surface of the earth that correspond roughly to latitude. The magnetic compasses of birds and sea turtles respond not to the polarity of the Earth's magnetic field, but instead to the inclination of magnetic field lines. Sensitivity to the inclination of the field allows animals to distinguish pole-ward from equator-ward. Migratory birds that cross the magnetic equator (flying in an equator-ward direction to get there) must therefore change their bearing to pole-ward in order to continue on their migratory path (Wiltschko and Wiltschko, 2002).

A variety of mechanisms have been proposed by which animals might detect magnetic fields, including magnetic induction, chemical magnetoreception, and the use of biogenic magnetite. There is no clear answer to how animals detect and orient to magnetic fields, but functional considerations dictate that whatever the mechanism may be, it must be sensitive to the inclination of the magnetic field (Lohmann and Johnsen, 2000). Behavioral experiments have set further functional criteria that the magnetic receptor must satisfy, such as sensitivity in the range of 10–50 nT (Walker *et al.*, 2002) against the background of the earth's magnetic field that ranges from 20 000–50 000 nT.

Kramer (1953) introduced the idea that oriented movement toward a goal has two distinct components, which he called the "map" and "compass." The map component refers to determining the relation of the animal's current position to the goal while the compass component consists of orienting in the correct direction to reach the goal. For animals to extract both map and compass information from the magnetic field of the earth requires sensitivity to different components of the magnetic field. This is because inclination is correlated with longitude and makes it possible for a migratory bird, as described above, to discriminate the pole-ward direction from the equator-ward direction (Wiltschko and Wiltschko, 2002). Obtaining positional information, however, requires discriminating the properties of the magnetic field at different locations (Walker *et al.*, 2002).

Among current proposals for magnetic sensory receptors are photopigment arrays in the retina of birds (Wiltschko and Wiltschko, 2002), and chains of membrane-enclosed magnetite crystals attached by microtubules to mechanically gated ion channels that open and close with oscillations of the magnetite chain (Walker *et al.*, 2002). Recently, magnetite chains have been described in the trigeminal afferents of the pigeon with properties that correspond well to those predicted by theoretical models (Fleissner *et al.*, 2003).

Although the exact nature of the magnetic sense and magnetic receptors remains tantalizingly unknown, many of its design features have been identified on the basis of known functions that this system is capable of performing. These functional specifications come from an understanding of the geophysical properties of the magnetic field of the earth and extensive behavioral research with homing pigeons, migratory birds, sea turtles, fish and other animals showing how they respond to experimental alterations of the local magnetic field.

### THE SYNTHESIS OF CAUSE AND FUNCTION

Although research on the cause and function of behavior are two different enterprises, with different aims and different practical methodologies, I have tried to show that ideas derived from considering the function of behavior have a number of important implications for the study of causation. Function defines the categories of behavior about which we ask causal questions, it can raise novel causal questions for examination, and it sets the criteria that satisfactory causal explanations must meet. The distinction between cause and function is nevertheless a crucial one and failing to distinguish the two can lead to confusion, faulty explanations of behavior, and fruitless debate (Hogan, 1994). I have been able find only one ethologist who made the radical suggestion that it might be possible to synthesize the study of cause and function into one seamless enterprise. He wrote:

"The fact that we tend to distinguish so sharply between the study of causes and the study of effects is due to what we could call an accident of human perception. We happen to observe behavior more readily than survival, and that is why we start at what is really an arbitrary point in the flow of events. If we would agree to take survival as the starting point of our inquiry, our problem would just be that of causation; we would ask: "How does the animal – an unstable 'improbable' system – manage to survive?" Both fields would fuse into one: the study of the causation of survival. Indeed, logically, survival should be the starting point of our studies. However, since we cannot ignore the fact that behavior rather than survival is the thing we observe directly, we have, for practical reasons, to start there. But this being so, we have to study both causation and effects."

It is a remarkable passage, and as the reader may have guessed, it comes from Tinbergen himself writing in the "Aims and Methods" paper. By "effects" he means what we would today call function. In Tinbergen's view there is a straightforward sequence of causal steps that ultimately affects survival, selection, and the evolution of behavior. For practical reasons, however, we have to dichotomize this sequence into the functional consequences of the behavior we observe and its causal antecedents. Because the goal of causal analysis, in his view, is to understand survival and the function of behavior he would not find it surprising that function defines the categories of behavior we investigate, raises new causal questions, and ultimately sets the criteria that causal explanations must satisfy.

### SUMMARY AND CONCLUSIONS

One of Tinbergen's most lasting contributions to the study of behavior was the distinction he drew among causal, functional, developmental, and evolutionary questions about behavior. More recently, behavioral ecologists have claimed that understanding the function of behavior is an important step in understanding its causes. This claim has, in turn, been criticized for confusing the fundamental distinction that Tinbergen articulated. The study of behavior, however, usually begins by identifying units of behavior functionally and only then proceeds to causal analysis. Research on four phenomena, disassortative mating by MHC loci, memory for cache sites by food-storing birds, auditory localization of prey by barn owls, and magnetic orientation illustrate the contributions to causal research of understanding the function of behavior. Understanding function – and sometimes simply a hypothesis about function – defines the causal questions that are asked, identifies novel questions for causal investigation, and sets the criteria that adequate causal explanations must meet.

### ACKNOWLEDGMENTS

I would like to thank Scott MacDougall-Shackleton and Mart Gross for valuable discussion, and Johan Bolhuis, Simon Verhulst, and Jerry Hogan for their comments on the manuscript. Many thanks to Johan

Bolhuis, the Royal Dutch Zoological Society, the Dutch Society for Behavioural Biology, and the Dutch Foundation for Scientific Research for the invitation to participate in the symposium marking the 40<sup>th</sup> anniversary of the publication of Tinbergen's influential paper. Preparation of this chapter was supported by the Natural Sciences and Engineering Research Council of Canada and by the Academic Development Fund of the University of Western Ontario.

JOHAN J. BOLHUIS

# 9

# Function and mechanism in neuroecology: looking for clues

In his groundbreaking paper, Tinbergen (1963, this volume) proposed that the four main problems in behavioral biology concerned causation, function (survival value), development (ontogeny) and evolution. Roughly speaking, causation and development are essentially causal or "how" questions, while function and evolution are functional or "what for" questions (cf. Hogan, 1994; Bolhuis and Macphail, 2001; Bateson, 2003; Bolhuis and Giraldeau, 2005). Functional or "what for" questions have also been called "why" questions (e.g., Bateson, 2003; Bolhuis and Giraldeau, 2005). This is confusing, as Tinbergen's four questions are also known as "the four whys"; for this reason I will use the term functional or "what for" questions. When Tinbergen wrote his paper, the prevailing emphasis in ethology was on causation. In his article he argued that more effort should be directed to the study of development, function, and evolution. Elsewhere in this volume, Crews and Groothuis and Hogan and Bolhuis discuss how developmental behavioral biologists took up Tinbergen's challenge. With regard to function and evolution, the behavioral ecology revolution in the 1970s was a somewhat delayed response to Tinbergen's clarion call. Recently, there have been increasing attempts to integrate the four questions. For example, in behavioral ecology there have been calls for a renewed research effort into the mechanisms of behavior (e.g., Krebs and Davies, 1997), and nowadays there is a field of research that is called – somewhat confusingly – functional ecology. In addition, many researchers studying cognitive and neural mechanisms increasingly appeal to the evolutionary history or the functional significance of behavior to better understand its causation (e.g. Kamil, 1998; Shettleworth, 1998; DeVoogd and Székely, 1998; Healy and Braithwaite, 2000; Sherry, 2005, this volume). This has led to a number of new fields of research, dubbed "evolutionary psychology," "cognitive ecology," and "neuroecology," respectively.

*Tinbergen's Legacy: Function and Mechanism in Behavioral Biology*, Johan J. Bolhuis and Simon Verhulst (eds.). Published by Cambridge University Press. © Cambridge University Press 2008.

In all three cases, functional or evolutionary considerations are used in an attempt to understand or explain the cognitive or neural mechanisms underlying behavior. At face value, this seems to be a laudable approach. All of Tinbergen's four questions are important, so they should all be addressed when investigating a particular behavioral problem. Hogan (1994) has pointed out, however, that cause and function are logically distinct categories. Bolhuis and Macphail (2001), following Hogan (1994), suggested that problems of mechanism (i.e., causation and development) cannot be solved by functional considerations (i.e., evolution and function). Shettleworth (1998, p. 574) has attempted to integrate causal and functional approaches to animal cognition, in what she called a "deep synthesis" of the "ecological program" and the "anthropocentric, traditional or general-process program." The author explicitly asked the question, "Can function explain mechanism?" (p. 42). From the remainder of her book it would seem that the answer to that question is affirmative. Although she says that "Throughout this book we will be concerned with the adaptive value and evolution of cognitive mechanisms," at the same time "In terms of Tinbergen's four questions, cognition – defined as information processing – is one of the proximate causes of behavior" (p. 13). Not surprisingly, she suggests that "(...) animal intelligence should ultimately be defined in terms of *fitness*" (p. 10). In contrast to Shettleworth's "deep synthesis," Bolhuis and Macphail (2001) maintained that "(...) it should be clear that functional questions and questions about mechanisms are fundamentally different, and furthermore, that results from one domain cannot be used as explanations in the complementary domain. Thus, for example, a functional interpretation of why an animal performs a certain behavior does not explain the cognitive and neural mechanisms governing that behaviour" (p. 426). We do acknowledge that functional or evolutionary considerations may provide clues as to a possible causal analysis of mechanisms, but add that these clues can be misleading as they do not necessarily reveal the actual mechanisms involved.

There have been a number of replies to the critique by Bolhuis and Macphail (2001). In this chapter I will address some of the main issues put forward in the replies (see also Bolhuis and Macphail, 2002), as well as some misunderstandings that have arisen. In addition, I will discuss recent evidence that is relevant to the debate. I will begin with a consideration of some clues provided by the 'maestro' (as Tinbergen was known by his Oxford students; see Kruuk, 2003) himself as regards the integration of the four whys.

What did Tinbergen say about the relationship between the four questions? First, he suggested that there was not a strict separation between them, and the distinction between the four problems was not necessarily logical. In his 1963 paper he wrote:

> " I admitted above that in speaking of 'the four problems of biology' we apply a classification of problems which is pragmatic rather than logical".
> (Tinbergen, 1963, p. 426 (see also Chapter 1, this volume))

Some authors have argued that the development of behavior is essentially a causal problem (e.g., Hogan, 1988; Bolhuis, 1999), but it clearly also has functional aspects (e.g., Oppenheim, 1981; see also Hogan and Bolhuis, this volume). Also, behavioral ecologists have argued that the study of mechanisms is essential for a proper understanding of the function of behavior (Krebs and Davies, 1997). So, it seems that Tinbergen was right in suggesting that the distinction between the four problems is not logical but pragmatic. However, as I indicated above, the four questions can be divided into predominantly causal (causation and development) and predominantly functional questions (function and evolution). Thus, among the four problems there are two types of questions that can be asked, namely causal and functional questions. Hogan (1994) has argued that cause and function are logically independent concepts. Thus, although Tinbergen was quite right when he suggested that the classification of the four problems is pragmatic, it is important to realize that causation and function are logically independent concepts that should not be confounded. We will return to the issue of the logical distinction between causation and function later in this chapter.

Second, Tinbergen stated that the four questions need to be distinguished, and that they are all equally important, and, third, that ethology has to give attention to their integration.

> "There is, of course, overlap between the fields covered by these questions, yet I believe with Huxley that it is useful both to distinguish between them and to insist that a comprehensive, coherent science of Ethology has to give equal attention to each of them and to their integration". (Tinbergen, 1963, p. 411)

We have already discussed the overlap that exists between the four questions, and many more examples could be provided. Most behavioral biologists would agree that all four questions are important for a full understanding of animal behavior. In the 1963 paper, Tinbergen discusses the four questions separately, and he does not say much about how they could be integrated. There are a few hints as to what he had in mind when he talked about integration, for instance when he writes that:

> " Teleology (...) can be said to have ceased to be a source of confusion in its cruder forms, in which function was given as a proximate cause (...)".
> (Tinbergen, 1963, p. 413)

In other words, Tinbergen suggests that function as a proximate cause has led to confusion. It would appear that he applauded the demise of this crude form of teleology. In fact, he implies that confusing cause and function is wrong, but at any rate is a thing of the past. Another hint as to the nature of the integration of the different questions is given in the following quote:

> "Zoophysiology [...] derives much of its inspiration and guidance from knowledge or hunches about survival value". (Tinbergen, 1963, p. 423)

"Zoophysiology" clearly refers to the study of causation, of the mechanisms underlying behavior. Tinbergen is rather cautious about the integration of causation and function here, when he says that the study of causation can "derive inspiration and guidance" from function, either from knowledge or from "hunches" concerning function.

### INTEGRATING FUNCTION AND MECHANISM

Integration of the four questions in the investigation of a particular behavioral problem is uncontroversial. Tinbergen and his collaborators have done that on several occasions, most notably in their famous study of egg shell removal in black-headed gulls, *Larus ridibundus* (Tinbergen *et al.*, 1962). Integration in this sense means investigating several of the four questions with respect to the same behavioral problem. As we have seen, Tinbergen (1963) was cautious concerning a conceptual integration of causation and function. More recent investigators have been more assertive concerning an integration of causal and functional questions. In fact, there is an increasing tendency to integrate Tinbergen's four questions in the study of brain and behavior, particularly in the new fields of evolutionary psychology, cognitive ecology, and neuroecology. These fields are part of a more general trend in biology, which has been termed the "ecological" "adaptationist" or "synthetic" approach (Kamil, 1988; Shettleworth, 1998).

Evolutionary psychology attempts to analyze human cognition and the underlying brain mechanisms using evolutionary considerations (e.g., Cosmides and Tooby, 1994; Daly and Wilson, 1999, 2005; Duchaine *et al.*, 2001). According to the most extreme version of evolutionary psychology, the human mind and brain can only be understood properly if selection pressures acting on our pleistocene hunter–gatherer ancestors are taken into account (Cosmides and Tooby, 1994; Duchaine *et al.*, 2001; cf. Laland and Brown, 2002; Buller, 2005). Real (1993) introduced the term "cognitive ecology," to indicate the adaptationist approach to the study of animal cognition. More recently, Healy and Braithwaite (2000, p. 22) described cognitive ecology as: "(…) work that attempts to integrate functional explanations of behaviour, such as those provided by behavioural ecology, with an understanding of the underlying psychological and neural mechanisms." Bolhuis (2000, p. 89), characterized some recent research into food storing and song in birds as "a 'neuroecological' approach to the study of the neural mechanisms of learning and memory." He stated that: "In a neuroecological approach, a functional or evolutionary principle is used for a comparative analysis of brain mechanisms of behavior (for some recent examples, see Balda *et al.* 1997; Shettleworth, 1998). For instance, by comparing the brains of a food storing bird species with those of a closely related species that does not store food, it is hoped that the neural substrate for the memory of stored food items will be revealed (cf. Krebs *et al.*, 1989)." In their critique of the adaptationist approach to the study of the brain mechanisms of learning and memory, Bolhuis and Macphail (2001) define "neuroecology" as: "The study of the neural mechanisms of behaviour guided by functional and evolutionary principles" (p. 426). The most extreme version of neuroecology was formulated recently by Healy *et al.* (2005), when they described it as "This approach of linking the size of specific regions of the brain to adaptations in behaviour and cognition" (p. 17).

## A CRITIQUE OF FUNCTIONAL APPROACHES TO COGNITIVE NEUROBIOLOGY

Macphail and Bolhuis (2001) criticized the functional or adaptationist approach to the study of brain and cognition. We contrasted an interpretation of cognitive evolution in terms of adaptive specializations with a general process view. Darwin (1871) had argued that between-species differences in cognitive capacities ("intelligence") were differences of degree, not of kind. An adaptationist or ecological approach to animal cognition proposes that animals have evolved species-specific and problem-specific mechanisms to solve problems associated with their particular ecological niches. Thus, different species have evolved different mechanisms, and within a species, different mechanisms are used to process information from different inputs. We argued that this adaptive specialization approach contrasts both with Darwin's view and with a general process view, according to which the same central processes of learning and memory are used across an extensive range of problems involving very different inputs. The adaptive specialization view predicts that, for instance, food storing birds should have evolved a specialized "spatial module(s)" (Shettleworth, 1998). As a result, food storers should be superior to non-storers in spatial memory tasks, but not in other memory tasks. After a detailed review of extant data on food storing birds, we found no consistent evidence to support the suggestion that, in the course of evolution, there has been adaptive specialization of spatial memory in food-storing birds.

Bolhuis and Macphail (2001) discuss a large number of examples, mainly from the neuroecological analysis of food storing and song learning in birds, that illustrate that functional or evolutionary clues may be misleading and often lead to the wrong answer with regard to the underlying neural mechanisms. We identified three ways in which these two key neuroecological paradigms are problematic, namely with regard to: (i) Localization of brain regions that are thought to be important for the behavior in question; (ii) Failures of behavioral predictions as to specialization of learning and memory capacities; (iii) Causal links between correlated behavioral and neural differences. I will summarize these three problems, using examples taken from bird song learning and food storing (for a detailed discussion see Bolhuis and Macphail, 2001 and Macphail and Bolhuis, 2001).

### (i)    Localization

Until recently, the neural substrate underlying bird song was thought to involve two forebrain pathways, the caudal and the rostral pathway (see Fig. 9.1). Within this so-called "song control system" (Clayton, 2000) – or simply "song system" (Nottebohm, 2000) –, the caudal pathway was thought to be involved in song production, while the rostral pathway was suggested to play a role in song learning. Evidence supporting the latter suggestion came from studies showing that lesions of parts of the rostral pathyway had relatively little impact on song in adults, while disrupting song acquisition in young males (see DeVoogd, 1994; Clayton, 2000; Nottebohm, 2000; Bolhuis and Eda-Fujiwara, 2003,

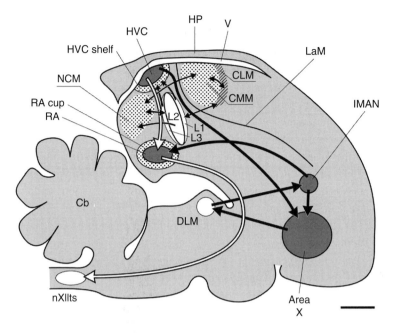

Fig. 9.1. Schematic diagram of a composite view of parasagittal sections of a songbird brain, with approximate positions of nuclei and brain regions. Lesion studies in adult and young songbirds led to the distinction between a caudal pathway (white arrows), considered to be involved in the production of song, and a rostral pathway (thick black arrows), thought to play a role in song acquisition (see text for further details). Thin black arrows indicate known connections between the Field L complex, a primary auditory processing region, and some other forebrain regions. The dark grey nuclei show significantly enhanced expression of immediate early genes (IEGs) when the bird is singing. Stippled areas represent brain regions that show increased IEG expression when the bird hears song, including tutor song. Abbreviations: Cb, cerebellum; CLM, caudal lateral mesopallium; CMM, caudal medial mesopallium; DLM, nucleus dorsolateralis anterior, pars medialis; DM, dorsomedial nucleus of the midbrain nucleus intercollicularis; HP, hippocampus; HVC, Acronym used as a proper name; formerly known as High Vocal Centre; L1, L2, L3, subdivisions of Field L; LaM, lamina mesopallialis; IMAN, lateral magnocellular nucleus of the anterior nidopallium; mMAN, medial magnocellular nucleus of the anterior nidopallium; NCM, caudal medial nidopallium; NIF, nucleus interface of the nidopallium; nXIIts, tracheosyringeal portion of the nucleus hypoglossus; RA, robust nucleus of the arcopallium; RAm, nucleus retroambigualis; rVRG, rostro-ventral respiratory group; V, ventricle; X, Area X. Scale bar represents 1 mm. (Adapted by permission from Macmillan Publishers Ltd: *Nature Reviews Neuroscience*, Bolhuis and Gahr, 2006, copyright 2006.)

for reviews). Identification of nuclei in the "song system" as the possible neural substrate underlying song learning was the result of a neuroecological approach where functional differences – between song-learners and non-song-learners, or between males and females, or between songbird males in the spring and in the autumn – were associated with neural differences. However, since the introduction of the immediate early gene (IEG) technique to this field (Mello *et al.*, 1992), it has become clear that brain regions outside the "song system" play an important role in song perception and possibly memory (Bolhuis and Eda-Fujiwara, 2003). In particular, exposure of adult zebra finch (*Taeniopygia guttata*) males to song leads to increased IEG expression in a number of regions outside the "song system," including the caudomedial nidopallium (NCM) and the caudomedial mesopallium (CMM; see Fig. 9.1).

Similarly, functional differences between food-hoarding bird species and non-hoarders have been associated with apparent differences in the volume of the avian hippocampus (e.g., Sherry *et al.*, 1989; Krebs *et al.*, 1989; for reviews see Clayton and Krebs, 1995a; Smulders and DeVoogd, 2000). Bolhuis and Macphail (2001) pointed out that, although the avian and mammalian hippocampi may be homologous, there are substantial differences between the two brain regions. Furthermore, neural homology does not imply homology at the level of cognition or behavior, and in fact the role of the mammalian hippocampus in memory is far from clear (cf. Murray, 1996; Brown, 2000; Buckley and Gaffan, 2000). Given this uncertainty, concentrating on the avian hippocampus as a possible substrate of spatial memory – ignoring the rest of the brain – is arbitrary.

### (ii)    Failure of behavioral predictions

Behavioral predictions resulting from an adaptationist approach have been formulated most clearly in the case of food storing in birds. Briefly, according to the adaptive specialization view, food-storing birds should be superior to non-storers specifically in spatial memory tasks. Performance in non-spatial memory tasks should be similar. Shettleworth (2003, p. 115) recently reiterated this prediction when she stated that "(…) an attractive and plausible hypothesis is that food storing birds possess an adaptation specific to spatial memory." Macphail and Bolhuis (2001) reviewed the relevant literature concerning this prediction, and found no evidence to support it. For example, four studies have employed various analogs of the radial maze, a much used test of spatial memory in rats. Hilton and Krebs (1990) found that storing parids were not superior to their non-storing counterparts. Three reports compared corvid species that differ markedly in the degree to which they rely upon stored food, and have not found a consistent association between degree of reliance on stored food and performance (Balda *et al.*, 1997; Kamil *et al.*, 1994; Gould-Beierle, 2000). Gould-Beierle (2000) found that Pinyon jays (who have a high dependence on stored food) and scrub jays (who have a much lesser dependence) performed significantly better than both jackdaws (*Corvus monedula*; who do not store food) or Clark's nutcrackers (who have the highest dependence of any corvid on stored food).

### (iii)   Correlations and causal links

Bolhuis and Macphail (2001) argued that there is no consistent relationship between hippocampus volume and food-storing category in birds. Recently, Shettleworth (2003, p. 110) acknowledged that "comparisons of spatial memory and hippocampus in food-storing and non-storing birds do not unanimously support the simple adaptationist hypothesis. For instance, there are exceptions to a simple linear relation between relative hippocampal volume and amount of storing thought to occur in the wild [e.g. Basil *et al.*, 1996]."

Sherry and Vaccarino (1989) found that aspiration lesions of the hippocampus of food-storing black-capped chickadees (*Poecile atricapilla*) impaired retrieval of stored food. Such lesion effects, of course, do not demonstrate a causal link between the hippocampus and spatial memory. Furthermore, when Sherry and Vaccarino tested chickadees in a task in which food was associated with a visual cue – a non-spatial task –, birds with hippocampal lesions made many more re-visits to incorrect cues than did controls. Similarly, Colombo and Broadbent (2000) reviewed evidence of disruption of memory in non-spatial tasks after hippocampal lesions in non-storing species.

Even if correlations between hippocampus volume and food storing are found, we cannot be sure of the causal relationship between these parameters. On the one hand, there may be similar correlations involving other brain regions. On the other hand, it is not clear which aspect of the animals' behavior is the relevant factor for such correlations – it is routinely assumed that this must be spatial memory, but there are many other factors that may be involved, such as stimulus perception or spatial navigation. As Tinbergen (1963) put it, "Yet it is, of course, true that in such 'natural experiments' one does not control the feature studied – one never knows which unknown aspect of the animal may have varied with the character studied" (p. 419).

Healy *et al.* (2005) explicitly included changes in volume of brain regions in their definition of neuroecology. With regard to food storing animals, they suggested that "the increased size of a particular brain region is associated with enhanced spatial memory" (p. 17). Surprisingly, no theoretical underpinning is provided for this bold assumption. Is "enhanced memory" associated with regional increases in brain size? Are neuronal processes increased in size or number? Are new neurons needed for new memories? Both of these suggestions have been made by Nottebohm (1981, 2002) as possible explanations of seasonal changes in neurogenesis and the volume of song control nuclei in songbirds. The most important question here is whether there is any evidence to support the thesis that changes in the size of brain regions are related to learning and memory. On the basis of evidence reviewed in Bolhuis and Macphail (2001) and in the present chapter, I maintain that the answer to this question is negative.

## SOME MISUNDERSTANDINGS IN THE NEUROECOLOGY DEBATE

A number of misunderstandings have arisen that threaten to cloud the debate concerning neuroecology. I will briefly discuss the major issues.

### Neuroecology is not neuroethology

The concept of neuroecology should not be confused with the established discipline of neuroethology. Neuroethology is the neurobiological analysis of animal behavior, using ethological methods and involves all four questions suggested by Tinbergen (1963, this volume). In contrast, neuroecology involves the study of the brain using functional concepts, as they used in, for example, behavioral ecology. The essence of the neuroecological approach is the idea that specification of the adaptive value of a behavior (investigated primarily by field observations) will in turn allow us to determine important features of the neural mechanisms underlying that behavior. This approach contrasts with the traditional neuroethological approach, which attempts to specify the neural underpinnings of behavior by neurobiological means. Ewert (1997), for example, in an exploration of the neural bases of key stimuli, does not appeal to functional aspects of prey recognition in trying to select between two alternative ways in which recognition might be achieved neurally. Instead, he analyses the responses of cells in the visual pathways involved.

## The importance of evolution

It has been suggested – implicitly or explicitly – that in their critique of neuroecology, Bolhuis and Macphail (2001) somehow belittle the role of evolution in behavior, or that these authors wish to discard functional or evolutionary considerations altogether. In fact, we explicitly stated that we feel that the study of animal behavior should involve all of Tinbergen's four questions (Bolhuis and Macphail, 2001, p. 426). Much of the confusion stems from the suggestion that evolution can be seen as a causal factor in behavior and cognition. For instance, Hampton *et al.* (2002, p. 6) stated that "The neuroecological approach begins by hypothesizing that evolution has shaped cognition." More recently, Healy *et al.* (2005) suggested that Macphail and I have criticized neuroecology for "confusing proximate and ultimate mechanisms." The expression "ultimate mechanisms" (see also Pravosudov and Clayton, 2001) again suggests a role for evolution or natural selection as a cause for behavior or cognition. Following Hogan (1994), we pointed out that results from the domain of mechanisms cannot be used as explanations in the domain of function and vice versa. In addition, Bolhuis and Macphail (2002) explicitly stated that "We do not distinguish between 'ultimate causes' and 'proximate causes' of behavior and cognition: there are no ultimate causes" (p. 8). Evolution works by means of natural selection. Natural selection operates on phenotypic variation. As such, natural selection might be said to be a causal factor in the historical *process* of evolution. As a result of natural selection, some phenotypes survive at the expense of others. However, it is important to realize that natural selection is *not* a causal factor for the (behavioral, neural or cognitive) phenotype. These phenotypes have their own causal factors, and these causal factors can be studied experimentally (cf. Hogan, 1994).

## The concept of modularity

In their commentary on Bolhuis and Macphail (2001), Flombaum *et al.* (2002) address the issue of modularity. In particular, they argue that "(…) neurobiological data cannot be used to bring a case against the thesis of modularity, whether in the case of memory and learning, or otherwise." Our paper was meant to be a critique of neuroecology, not of the concept of modularity. Many researchers using the neuroecological approach do not use the notion of modularity at all, so it is not a crucial concept. However, some neuroecologial investigators do make statements about modularity (e.g., Pinker, 1997; Shettleworth, 1998, 2000) and in our critique we wished to point out that their view of modularity is different from Fodor's (1983) original suggestions, which we outlined briefly. Nonetheless, because Flombaum *et al.* (2002) suggest that we have misinterpreted Fodor's (1983) concept of modularity, I shall briefly discuss the three issues they raised.

### Are modules "innate?"

Flombaum *et al.* (2002) suggest that there is no reason to assume that they are, and in fact, so they propose, often they are not. Fodor (1983) stated explicitly that modules are "innately specified," however, and we merely repeated what he said. Pinker (1997) suggests that "Saying that the different ways of knowing [i.e., modules] are innate is different from saying that knowledge is innate" and "Talking about innate modules is not meant to minimize learning but to explain it" (p. 315). It would seem that for a neuroecological interpretation, somewhere along the line there has to be something about "adaptively specialized modules" that is heritable, otherwise these modules could not have been adaptively specialized through an evolutionary process. As Bolhuis and Macphail (2001) do not believe that learning and memory mechanisms are adaptively specialized modules, this is really a problem that should be addressed by those who follow a neuroecological approach. For a recent critical discussion of the concept of "innate," see Bolhuis (2005).

### Are modules domain-specific?

According to Flombaum *et al.* (2002) they "need not be." Fodor (1983) suggested that they are. More recently (Fodor, 2000) returned to this issue and in his characteristic style suggested that "It's often said that what makes something a module is that it's "domain specific." There's a point to this, I think, but it needs to be handled with care." (p. 58). In a subsequent analysis ("I'm afraid that this is going to seem a bit scholastic") he suggests that domain specificity does not apply to process per se, but *"Rather, it applies to the way that information and process interact"* (p. 60, his italics). Pinker (1997) suggested that informational encapsulation is not a defining characteristic of modules. He seems to suggest that domain specificity may be, when he states that "Modules are defined by the special things they do with the information available to them, not necessarily by the kinds of information they have available" (p. 31). In contrast, Fodor (1983, 2000) suggested that informational encapsulation is what makes modules modular. Presumably, domain specificity is what makes modules "specialized." Bolhuis and Macphail (2001) did not find consistent evidence for specialized cognitive systems for learning and memory, so again this would seem to be a problem for neuroecology, not for its critics.

### Modularity of the brain?

Flombaum *et al.*'s main criticism of our paper can be summarized with their statement: "It is not, however, the case that a psychological module must have a corresponding neural module in the sense of a specified brain nucleus" (p. 108). This criticism is specifically directed at Box 1 of Bolhuis and Macphail (2001) where we state that, according to Fodor (1983), modules are "hardwired, and located in specific brain regions." This would seem to be a fair representation of Fodor's (1983) views, since he clearly states that he suspects modular systems to be "hardwired" which means that they are "associated with specific, localized, and elaborately structured neural systems" (p. 37). Later he suggests that "there is a characteristic [fixed] neural architecture associated with each of what I have been calling the input systems" (p. 98). It would seem logical – and indeed crucial – for researchers applying the modularity concept to neuroecology to assume that modules are "hardwired," and that cognitive modularity is accompanied by neural modularity. This is exactly what neuroecologists do. For instance, Shettleworth (2003) states: "(…) an attractive and plausible hypothesis is that food storing birds possess an adaptation specific to spatial memory and its neural substrate. Many of the data resulting from attempts to test this hypothesis are consistent with it." (p. 115). Another example is provided by Healy *et al.* (2005) who explicitly defined the neuroecological approach as involving "linking" behavioral or cognitive adaptations to "the size of specific regions of the brain" (p. 17).

Neuroecologists may not always be so explicit about neural modularity, but investigators using this approach usually focus on a brain region (or brain regions) that is thought to be crucial for the learning process involved, and a likely substrate for the particular memory. For instance, in investigations of food storing birds it is usually assumed that the hippocampus is the neural substrate of information stored in spatial memory. For example, Smulders and DeVoogd (2000)'s review is entitled

"The avian hippocampal formation and memory for hoarded food: Spatial learning out in the real world," in which they state that the structure "(…) plays an important role in processing this form of spatial memory" (p. 127). In another example, Krebs *et al.* (1996) call their review "The ecology of the avian brain: food storing memory and the hippocampus." Clayton and Krebs (1995) state: "Thus, memory in food-storing birds provides the first documented example of an evolutionary specialization of the brain which is associated with memory processing (…)" (p. 139–140), and: "The idea is that the ecological niche of utilising scatter-hoarded food for winter survival has generated a selection pressure for certain special properties of memory, which in turn has led to the evolution of specialised brain structures associated with processing these memories" (p. 140). Healy and Krebs (1992) state: "(…) it has therefore been hypothesized that the enlargement of the hippocampal region of food-storing birds represents an adaptation associated with a specialized memory capacity for retrieving stores"(p. 241). Finally, Sherry *et al.* (1989) state: "The greater size of the hippocampus in food-storing birds provides a clear case in which natural selection has modified a brain region involved in a cognitive component of behavior. The increased size of the hippocampus in 3 unrelated families of food-storing passerines indicates that natural selection favouring food storing has resulted in modification of the brain region that plays a central role in memory for cache sites." (p. 316). Likewise, in the study of bird song learning it is usually assumed that some of the so-called "song control nuclei" (particularly those in the rostral pathway; see Fig. 9.1) are the substrate of song memories (e.g., DeVoogd, 1994; see below for a discussion of the nature of song memory).

Lefebvre and Bolhuis (2003) have argued that significant positive correlations between food storing category and hippocampus volume would be consistent with neurocognitive modularity. Lefebvre and his colleagues have found positive correlations between avian innovation frequency and tool use and certain forms of learning. In addition, innovation rate in birds correlates significantly positively with the volume of the mesopallium/nidopallium complex (Lefebvre *et al.*, 2002; Timmermans *et al.*, 2000; Lefebvre and Bolhuis, 2003). In contrast, Lefebvre and Bolhuis (2003) reported a significant negative correlation (Eurpean parids, North American corvids) or no correlation (Eurasian corvids) between innovation frequency and the degree of food storing in birds.

The negative and zero correlations reported by Lefebvre and Bolhuis (2003) are consistent with the view that food-storing is modular, but they do not tell us which aspect of food storing (dead reckoning, landmark use (Shettleworth, 2000), navigation, working memory, long term memory, stimulus perception) constitutes the "spatial module," nor whether the module is central or restricted to the input level, as Fodor (1983) suggested. Innovations can be described as new ways to solve problems, and Fodor (1983) considers "problem solving" an operation performed by central systems. As Bolhuis and Macphail (2001) have argued, the modularity that has been claimed for food storing birds and songbirds may not be a modularity of memory systems, but of input systems. It may well be that positive correlations between food storing category and hippocampal volume are not due to variation in spatial learning and memory, but to variation in some kind of input system such as aspects of spatial navigation (Bolhuis and Macphail, 2001). Lefebvre and Bolhuis (2003) noted that innovation rate and learning (which might be called operations of central systems) show positive correlations with the volume of large sections of the brain or whole brain size. On the other hand, performance involving adaptively specialized "input systems" such as those involved in food storing, has been found to correlate with the volume of a restricted brain region, the hippocampus, and not with the volume of the telencephalon (e.g., Healy and Krebs, 1992; but see Garamszegi and Eens, 2004). This is broadly consistent with a Fodorian view of "hardwired" modularity restricted to input systems, with central systems being non-modular and having an "equipotential" (Fodor, 1983) representation in the brain. In food storing, these input systems could be those involved in spatial navigation or spatial stimulus processing. Differences in hippocampus volume between food storers and non-storers would then suggest a kind of parallel neural modularity akin to Fodor's "hardwired" modules. In conclusion, a significant positive relationship between food storing category and hippocampus volume would be consistent with neurocognitive modularity in the sense of Fodor (1983). However, following Fodor (1983), such modularity may not apply to central systems such as those involved in learning and memory, but to input systems such as those involved in spatial navigation.

### Neuroecology and scientific progress

A number of critics have suggested that Bolhuis and Macphail (2001) do not adequately appreciate the way in which science progresses. For instance, Hampton *et al.* (2002) suggested that "Neuroecology is subject to the difficulties inherent to any scientific endeavor." Demonstrating some feeling for melodrama, they further state that "It is certain that neuroecologists will often be misled, hypotheses will be discarded and refined, but the science of neuroecology will stagger forward" (p. 7). Similarly, MacDougall-Shackleton and Ball (2002)'s reply is entitled "Revising hypotheses does not indicate a flawed approach." These authors suggest that there is nothing wrong in revising a hypothesis after it has been disproved. This is true, and it is also the case that neuroecology has been useful in generating hypotheses as to the mechanisms of behavior. However, neuroecological hypotheses are not intrinsically superior to other types of hypotheses. If the basic axioma that there has been adaptive specialization of brain and cognition is wrong, then neuroecological hypotheses are doomed. In addition, on the basis of functional differences between species, neuroecologists look for correlations between arbitrary measures (e.g., volume) of arbitrary brain regions (e.g., the hippocampus) and arbitrary behavioral (e.g., food storing tendency) or cognitive (e.g., spatial memory) variables. This is an odd and potentially misleading empirical approach. On the one hand, the number of hypotheses that can be generated in this way is infinite – one could equally abritrary switch to other brain regions or other behaviors, as has been done (see below). On the other hand, restricting scientific investigation to an arbitrary brain region or cognitive mechanism is misleading.

Interestingly, disappointing results have led to revising hypotheses only in the case of birdsong, not in the food storing paradigm. The absence of a clear correlation between song acquisition and volume of "song control nuclei" (e.g., Brenowitz et al., 1991, 1995) has resulted in a focus on song production (MacDougall-Shackleton and Ball, 1999) or song stereotypy (Tramontin and Brenowitz, 2000) as the possible crucial factor. On the other hand, in recent research into song memory, focus has shifted away from the "song system" to brain regions outside it (Bolhuis & Eda-Fujiwara, 2003; see below). In contrast, failures to find a significant correlation between hippocampus volume and food storing category in birds (e.g., Brodin & Lundborg, 2003) has not led to revised or new hypotheses concerning the neural substrate of spatial memory. The only exception is Garamszegi and Eens (2004b), who, on the basis of a meta-analysis (see below for details), concluded that the volume of brain regions other than the hippocampus is related to food storing category.

### RECENT NEUROECOLOGICAL EVIDENCE

Since the publication of the critique of neuroecology by Bolhuis and Macphail (2001, 2002; Macphail and Bolhuis, 2001), there have been a number of new empirical papers on both song learning and food storing in birds, some of which are relevant to the neuroecology debate.

## Bird song memory; the song system and beyond

It has been proposed that the volume of certain nuclei in the song system of songbirds is related to the size of their song repertoire (e.g., Nottebohm *et al.*, 1981; DeVoogd *et al.*, 1993). Although a number of studies have found a significant relationship between particularly the volume of HVC (see Fig. 9.1) and song repertoire, both within and between species (e.g., Nottebohm *et al.*, 1981; DeVoogd *et al.*, 1993; Ward *et al.*, 1998; Airey *et al.*, 2000; Airey and DeVoogd, 2000;), there are also several studies that did not find such a relationship (e.g., Kirn *et al.*, 1989; Bernard *et al.*, 1996; MacDougall-Shackleton *et al.*, 1998; Leitner & Cathpole, 2004; for a review see Garamszegi and Eens, 2004). Recently, Leitner and Catchpole (2004) could not replicate Nottebohm *et al.*'s (1981) original finding of a correlation between syllable repertoire size and HVC volume in the canary (*Serinus canaria*), although they did find a postive correlation between the number of what they call "sexy syllables" and HVC volume. Brenowitz *et al.* (1991) also did not find a consistent relationship between seasonal variation in volume of song nuclei and repertoire size in a number of songbird species.

In a meta-analysis of published results, Garamszegi and Eens (2004) found a significant positive correlation between the volume of the forebrain nuclei HVC and Area X (relative to the volume of the telencephalon) and song length and repertoire size. When the covariation of song length and repertoire size was taken into account, the authors found that these two traits independently accounted for a significant amount of variation in the relative volume of the HVC, but not of Area X. Meta-analysis such as performed by these authors is not without its critics (e.g., Palmer, 2000; Jennions and Møller, 2002). In addition, it is difficult to interpret "effect sizes" in a number of different studies that varied in the methods used to estimate the volume of brain nuclei. Gahr (1997) found that different methods of measuring the volume of brain nuclei will give different results (see below). In any case, the meta-analysis by Garamszegi and Eens (2004) does not permit a conclusion as to the causal relationship between these variables. In an earlier study, Ward *et al.* (1998), had found significant positive correlations between number of song elements copied and HVC volume and neuron number in zebra finches. These authors reached a firm conclusion concerning the nature of the causal relation-ship, when they state that: "(…) naturally occurring variation in neuron number constrains how much song material can be copied or repro-duced." Whatever the nature of the causal relationship, correlations such as those reported by Garamszegi and Eens (2004) are not easy to interpret because it is not clear which aspect of the animals' behavior is crucial. It is sometimes suggested that this is song learning, but, alter-natively, variation in HVC volume could be associated with variation in song production (as MacDougall-Shackleton and Ball (1999) have sug-gested for sex differences) or with song stereotypy (as Tramontin and Brenowitz (2000) have suggested for seasonal variation). It is perhaps because of this uncertainty that Garamszegi and Eens (2004) speak of "neural correlates of singing behavior" in the title of their paper.

If volume of brain regions is a rather crude measure of neural mechanisms, it could be argued that measurements at the level of neurons or synapses are more appropriate. This is exactly what Nealen (2005) did, when he measured synapse density in two nuclei in the song system (HVC and RA; see Fig. 9.1) in males and females of the zebra finch and the Carolina wren (*Thryothorus ludovicianus*). Zebra finch males sing only one stereotyped song, while Carolina wren males have a large repertoire of more than 20 different songs. Females of both species do not sing. Despite large differences in song repertoire, syn-apse density in both HVC and RA was similar in males of both species. There were significant effects of sex on synapse density in both brain regions, but these were of a rather surprising nature. Synapse density in both HVC and RA of the female Carolina wren was comparable to that in males of both species. In contrast, synapse density in both regions in female zebra finches was significantly greater than in males of both species, as well as in female Carolina wrens. These findings do not suggest that there is a relationship between synapse density in these song-related nuclei and repertoire size, sex differences in vocal production or differences in song perception.

In research into the neural mechanisms of bird song learning there has been increasing interest in the role of brain regions outside the conventional song system, particularly the caudomedial nidopal-lium (NCM) and the caudomedial mesopallium (CMM; previously known as caudomedial hyperstriatum ventrale or CMHV; see Reiner *et al.*, 2004, for new avian nomenclature) as possible substrates for song memory (Fig. 9.1, for reviews see Clayton (2000), Mello (2002), Bolhuis and Eda-Fujiwara (2003), Bolhuis and Gahr (2006), and Bolhuis (2008)). Bolhuis *et al.* (2000, 2001) found a significant positive correlation between neuronal activation (measured as expression of protein products of immediate early genes) in the NCM and the strength of song learning (measured as the proportion of elements of the tutor song copied by the experimental males). Terpstra *et al.* (2004) confirmed the significant relationship between the strength of song learning and neuronal activation in the NCM in zebra finch males, and showed that this relationship is unlikely to be related to attentional processes. Recently, Gobes and Bolhuis (2007) reported that lesions to the NCM impaired recognition of the tutor song in zebra finch males, without affecting their song output. Taken together, these findings suggest that the NCM may be (part of) the neural substrate for the representation of tutor song memory in male zebra finches. Female zebra finches do not sing, but nevertheless they learn the characteristics of tutor song and form a preference for it (Riebel *et al.*, 2002). Terpstra *et al.* (2006) found that adult zebra finch females that were re-exposed to their father's song showed significantly greater IEG expression than controls that were exposed to a novel song in the CMM, but not in the NCM or hippocampus. These results suggest that in female zebra finches, the CMM may be (part of) the neural substrate for the representation of tutor song memory.

### Birdsong: One brain for all seasons

Seasonal variation in the volume of certain nuclei in the song system of songbirds has been an important neuroecological paradigm. This phenomenon was discovered by Nottebohm (1981), who suggested that the greater volume of the canary song nuclei HVC and RA (see Fig. 9.1) in the spring, is associated with the acquisition of new songs in the breeding season. Basically, learning of new songs was thought to require new neuronal processes enabling the formation of new synapses, which would lead to an increase in volume of the relevant brain regions. Whatever the nature of the learning that was thought to be associated with the seasonal volume changes, the hypothesis was not confirmed in subsequent research by Brenowitz and colleagues (e.g., Brenowitz *et al.*, 1991). These authors found that these seasonal changes are not related to the learning of new songs, song repertoire or song output. They suggested that the seasonal changes may be related to song stereotypy, with songs being more stereotyped (with less song-to-song variability) during the spring breeding season (Tramontin and Brenowitz, 2000).

Recent research by Gahr and his colleagues (Leitner *et al.*, 2001) has cast more light on this phenomenon. Previously, Gahr (1997) had shown that estimates of the volume of brain nuclei are strongly dependent on the neuroanatomical technique that is employed. Apparent seasonal changes in volume of the HVC, when measured in Nissl stained sections, cannot be found when sections are stained using neurochemical or retrograde labeling. Gahr (1997) suggested that there is no seasonal change in the number of neurons in HVC, but that the phenotype of permanent HVC neurons changes with season. Leitner et al. (2001) investigated the songs and brains of free-living wild canaries in a longitudinal field study on Madeira. These authors found that free-living canaries actually "recycle" approximately 50% of their repertoire in the following breeding season, while approximately 12% of the repertoire is actually new. Thus, song plasticity in free-living canaries is less dramatic than in their domesticated counterparts. Nevertheless, there is considerable seasonal plasticity in the syllable repertoire of free-living canaries. Surprisingly, there was no significant change in the volume of either HVC or RA in these birds, whether measured in conventional Nissl stained sections or as expression of androgen receptor mRNA. In addition, the number of neurons and neuron density was not significantly different between the breeding season and the non-breeding season. These results in wild canaries do not support the suggestion that there is a relationship between seasonal changes in the song repertoire and volume changes of song control nuclei.

There have been numerous reports of neurogenesis in nuclei in the song system of adult songbirds (e.g., Goldman and Nottebohm, 1983). Nottebohm and collaborators suggested that adult neurogenesis in songbirds is related to learning of new songs (Goldman and Nottebohm, 1983; Kirn et al., 1994; see Nottebohm, 2002, for a recent review). These findings have led Nottebohm (2002) to propose a new theory of the neural representation of memory, when he stated that "Those neurons, and not their individual synapses, would be the unit of long-term memory, and the number of neurons that can be modified in this manner would determine how much can be learned and remembered" (p. 627). In canaries, seasonal recruitment of new neurons in the HVC (Fig. 9.1) seemed to coincide with the acquisition of new song syllables to the repertoire. However, Tramontin and Brenowitz (1999) found that there was seasonal incorporation of new neurons in the HVC of the western song sparrow (Melospiza melodia morphna), a songbird species that does not show seasonal changes in song learning. Thus, in contrast to canaries, western song sparrows do not modify song syllables, nor do they add new syllables to their repertoire as adults. Nevertheless, these birds showed seasonal variation in incorp-oration of new neurons in HVC similar to canaries, prompting the authors to conclude that "The functional significance of neuronal recruitment into HVC therefore remains elusive." In a critical review, Gahr *et al.* (2002) are equally sceptical about the relationship between neurogenesis and learning and memory.

### Correlations between hippocampus volume and food storing category

A number of recent studies have addressed the possible correlation between relative hippocampus volume and food storing category in birds. Arguably the most important study is that by Brodin and Lundborg (2003), who published a correlation analysis of the largest sample of corvids and parids to date. Figure 9.2 shows an updated and amended version of their results, with a few species now with a differ-ent classification (see legend to Fig. 9.2). There was no significant correlation between food storing category and hippocampal volume in either corvids or parids in the original paper (Brodin & Lundborg, 2003) nor in the updated version of their results as shown in Fig. 9.2.

Recently, Lucas *et al.* (2004) reanalyzed the data by Brodin and Lundborg (2003). Surprisingly, these authors found that Eurasian cor-vids and parids had significantly larger hippocampi than their North American counterparts. The fact that North-American storers have smaller hippocampi than Eurasian ones (North-American corvids also have a smaller telencephalon and lower body mass than their Eurasian counterparts) cannot easily be explained by the authors, but they offer a number of suggestions. It is possible that the difference is a result of different laboratories treating animals and brains differently. In the case of corvids, the authors note that virtually all North American and Eurasian birds were analyzed by only two different laboratories. Alternatively, it may be that birds from the different continents were taken from different habitats. Pravosudov and Clayton (2002) found significant differences in hippocampus volume, volume of the rest of the telencephalon and body mass between black-capped chickadees from populations from different latitudes. Similar population differ-ences may have contributed to the continent effect reported by Lucas *et al.* (2004). From a neuroecological perspective, it is worrying that such an arbitrary factor has such a large effect on variance.

In a recent meta-analysis (Garamszegi and Eens, 2004b), which also includes non-passerines, surprisingly there was a significant

(a)

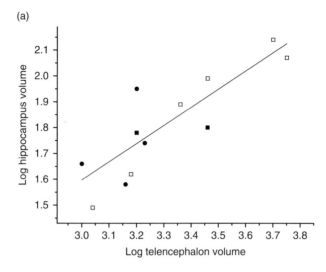

(b)

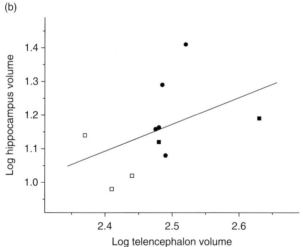

Fig. 9.2. Hippocampus volume (log mm³) plotted against telencephalon volume (log mm3) in corvids (a) and parids (b), in three categories (following Brodin and Lundborg, 2003 and Lucas *et al.*, 2004): non-storers (filled squares), category 2 or non-specialized storers (squares) and category 3 or specialized storers (filled circles). According to the definitions of Brodin and Lundborg (2003) category 2 species store small amounts of food during short time intervals while species in category 3 store thousands of food items in the autumn to be used as winter food. (Adapted from Brodin and Lundborg (2003) and Lucas *et al.* (2004), with permission.) Following Brodin and Lundborg (2003; Table 2), in the present figure, black-capped chickadee, coal tit (*Parus ater*) and marsh tit (*P. palustris*) were moved from category 2 to category 3 storers. (Figure courtesy of A. Brodin.)

positive correlation between food storing category and brain mass. The authors suggest that brain regions other than the hippocampus may be involved in spatial memory. Bolhuis and Macphail (2001) had already stated that it is not clear why neuroecologists focus on the hippocampus when it comes to food hoarding in birds: why is the rest of the brain ignored? I have already discussed the problems involved in meta-analyses in the context of song learning, above.

Pravosudov and Clayton (2002) reported differences in absolute and relative hippocampal volume between populations within the same species, the black-capped chickadee. Northern populations experience longer and colder winters than more southern ones. Assuming that this would be reflected in the tendency to store food, the authors compared hoarding behavior in a northern (Alaska) and a southern (Colorado) population of black-capped chickadees. The birds were transferred to a laboratory in the Autumn where they were tested under identical, controlled conditions. Chickadees from Alaska stored at higher rates, remembered caching locations more accurately, and had larger hippocampi than their conspecifics from Colorado. This is an interesting finding, prompting Shettleworth (2003) to suggest that future research in this field should focus on differences between different populations within the same species. However, neuroecology applied to within-species comparisons suffers from the same problems that between-species comparisons do. There are likely to be many more differences between the two chickadee populations than just food storing rates and spatial memory. And, just as in between-species comparisons, focusing on the hippocampus is arbitrary and potentially misleading.

These problems are illustrated by two recent reports. Pravosudov and Clayton (2001) had found that captive mountain chickadees (*Poecile gambeli*) kept on a limited and unpredictable food supply were more accurate at recovering their stored food and performed better in spatial memory tests than birds maintained on *ad libitum* food. Importantly, the two experimental groups did not differ in the amount of food that was stored. In a subsequent study using the same birds, Pravosudov *et al.* (2002) showed that – contrary to their prediction – there were no significant differences in volume or total neuron number of the hippocampus between the two experimental groups. The authors argue that the length of the feeding treatment (3 months) should have been sufficient to obtain detectable differences in the hippocampus. They conclude that there is not necessarily a relationship between accuracy of recovery of stored food and spatial memory performance on the one hand, and hippocampal volume or neuron number on the other. This is an important finding, as, unlike in the study by Pravosudov and Clayton (2002), the birds in these two groups did not differ in the amount of food stored. In addition, the difference in spatial memory performance between experimental groups in the study by Pravosudov *et al.* (2002) was obtained under controlled laboratory conditions, while Pravosudov and Clayton (2002) used wild-caught birds from two different populations that may have differed in all kinds of ways.

The problems involved in within-species comparisons are further illustrated by a field study conducted by Brodin (2005), who investigated whether relative hippocampus volume can be used as a predictor for food storing category. This author observed storing behavior of black-capped chickadees in British Columbia, and compared it with the behavior of the closely related Willow Tit, *P. montanus*, that he had observed previously in Sweden. Brodin explains that these two species used to be considered conspecifics and are often treated as a super-species, the *atricapilla* complex. Although the two species live under very similar ecological conditions, the hippocampus of the willow tit is almost twice the size of that of the black-capped chickadee. Brodin (2005) found that chickadees stored food at the same high rates as willow tits. Thus, in what amounts to virtually a within-species comparison, there is no clear relationship between food storing capacity and hippocampus volume in this field study. Brodin (2005) suggests that food storing in the laboratory may not always follow the pattern predicted by field observations.

Pravosudov *et al.* (2001) found that the unpredictable food supply such as used by Pravosudov and Clayton (2001) and Pravosudov *et al.* (2002) resulted in chronically elevated corticosterone levels in mountain chickadees. Subsequently, Pravosudov (2003) reported that mountain chickadees with experimentally increased corticosterone levels stored more food, and performed significantly better in spatial memory tasks than controls. However, in a recent study, Pravosudov and Omanska (2005) found that chickadees with such elevated corticosterone levels did not differ from controls in hippocampal volume, neuron number, and neurogenesis. It is likely that chickadees living in the relatively harsh conditions of Alaska experience more stress and have a more irregular food supply than their counterparts in Colorado (Pravosudov and Clayton, 2002), which may have led to enhanced corticosterone levels in the former. These studies demonstrate, however, that although an increased corticosterone level may be an intervening variable leading to increased food storing and spatial memory, it has no effect on hippo-campal morphology.

## Spatial memory in food-storing and non-storing birds

Shettleworth and Westwood (2002) attempted to further test neuro-ecological hypotheses of spatial memory in a food-storing (black-capped chickadees) and non-storing bird species (juncos, *Junco hyemalis*). Previously, Hampton and Shettleworth (1996a,b) had compared the same two species in a series of hippocampal lesion experiments. As Bolhuis and Macphail (2001) noted, the results of these two studies challenge the notion that a larger hippocampus should be associated with better performance in tasks that are susceptible to hippocampal damage. Hampton and Shettleworth (1996a,b) found that there are spatial memory tasks, susceptible to disruption by hippocampal lesions in both species, in which the performance of (intact) non-storing juncos is either no different from or actually superior to, that of intact chickadees. Thus, Hampton and Shettleworth (1996b) concluded that "sensitivity of a given task to hippocampal damage does not predict the direction of memory differences between storing and non-storing species" (p. 946). The junco hippocampus is relatively smaller than that of the black-capped chickadee. Hampton and Shettleworth's results support the suggestion that the hippocampus is involved in the spatial tasks that they employed. Thus, as Bolhuis and Macphail (2001) concluded, a relatively large hippocampus is not reliably associated with superior performance in spatial memory tasks.

In the abstract of their recent paper, Shettleworth and Westwood (2002) suggest that chickadees perform better in spatial memory tasks than in non-spatial memory tasks, compared to juncos (cf. Shettleworth, 2003). However, closer inspection of their results does not warrant this conclusion. The authors performed a series of delayed-matching-to-sample tasks with either the color or the spatial position of the stimuli on a computer screen having to be recognized in the test. In two different experiments, relative performance in the two memory tasks was meas-ured, using different retention intervals. In both experiments there was no significant effect of the factor Species (junco or chickadee), and neither were there any significant interactions involving the factor Species. In fact, the authors conclude that "In contrast to predictions based on previous experiments (...) both chickadees and juncos per-formed as if they processed the location of a stimulus at the expense of processing its color" (p. 232); and "Contrary to what their food-storing status predicts, however, the chickadees' location matching performance declined more as the RI [retention interval] increased than did their color matching" (p. 233). Thus, whether an absolute or a relative criterion of memory performance of storers versus non-storers is used, these results provide no evidence to support the neuroecological hypothesis of superior spatial memory in food-storers. Within the two groups, there was a significant effect of Stimulus type (spatial or color) in the chickadees but not in the juncos, which is taken by the authors as evidence supporting the suggestion that storers should be superior in relative spatial memory (relative to non-spatial memory). However, statistically there is no justification for within-species analysis in the absence of significant effects of Species or any significant interactions involving the factor Species. In addition, in both relevant experiments the juncos experienced significantly longer actual retention intervals than the chickadees, which leads the authors to state that "these data permit only tentative conclusions regarding possible species differences in memory" (p. 232). Finally, and most importantly, closer inspection of the results of Experiment 2 of Shettleworth and Westwood (2002) reveals that the trend of a differential effect of Stimulus type (spatial or color) was mainly due to the juncos' performance in the color task being superior to that of the chickadees; in the spatial task, performance in the two groups was similar. In addition, as the authors note, there was a significant decline in performance in the spatial task with retention interval in the chickadees, not in the juncos. In conclusion, the trend towards superiority of storers to non-storers in a relative measure of spatial cognitive performance seems to be due mainly to superior performance of the non-storers in the non-spatial "control" task.

## Seasonal variation in brain parameters in food storing birds and mammals

Black-capped chickadees store food items particularly in the Autumn. Barnea and Nottebohm (1994) reported that the number of newly recruited neurons in the hippocampus of wild-caught black-capped chickadees was significantly greater in October than in any other month of the year. Neuronal recruitment was defined as the number of labelled neurons in the hippocampus 6 weeks after injection of the cell birth marker [$^3$H] thymidine. It should be noted that the sampling distribution in this study was somewhat uneven. There were 10 data-points for October, but only 5 in total for the Winter months of

December and February, and only one datapoint each for the months of May, June, and July. In a critical discussion of these findings, Pravosudov and Omanska (2005) suggest that they do not establish a link between neuronal recruitment and spatial memory because (i) spatial memory performance is known to vary between individuals, and individual performance is unknown – this seems a valid point in view of the uneven distribution of sampled birds over the different months; (ii) There is insufficient data on the seasonality of food-storing behavior in wild chickadees (the authors suggest that there might be a second peak in food storing in early spring, as has been reported for European parids); (iii) It is not clear whether chickadees have better spatial memory when they are storing food (in October) or when they retrieve food much later.

Barnea and Nottebohm (1994) did not find seasonal variation in hippocampus volume or in the number of neurons in this brain region. In contrast, Smulders *et al.* (1995, 2000) reported seasonal changes in black-capped chickadees, with both hippocampus volume and neuron number being significantly greater in October (when the birds store food) than in January or June. Lavenex *et al.* (2002b) and Pravosudov *et al.* (2002) have discussed possible reasons for the discrepancy between the findings of Barnea and Nottebohm (1994) and of Smulders *et al.* (1995, 2000). They suggest that Smulders *et al.* (1995) have included juvenile birds in their sample, which may have affected their results. Lavenex *et al.* (2002b) concluded from a reanalysis of the data of adult birds of Smulders *et al.* (1995) that the volume of the telencephalon varied seasonally in their study. Lavenex *et al.* (2002b) found that the seasonal variation in hippocampus volume in the study of Smulders *et al.* (1995) closely followed the seasonal variation in the volume of the telencephalon, such that "these variations do not appear to be specific to the hippocampal formation." Interestingly, in an early publication of their results, Smulders *et al.* (1993) had reported that the volume of the telencephalon varied seasonally in chickadees, and that variation in hippocampus volume could be explained by this.

Recently, Hoshooley and Sherry (2004) reported a lack of seasonal variation in hippocampus volume, neuron number, neurogenesis, and cell death, in black-capped chickadees sampled in October, November, January, February, and March. In this study, the birds were sacrificed 2 weeks after injection of the cell birth marker 5-bromo-2-deoxyuridine (BrdU). Considering their findings together with the results of Barnea and Nottebohm (1994), Hoshooley and Sherry (2004) concluded that enhanced hippocampal neuronal recruitment in the Autumn is not due to increased neurogenesis, but rather a result of enhanced survival of new neurons. Interestingly, although these authors did not find seasonal variation in any hippocampal variable, the volume of one of their "control regions," the mesopallium/hyperpallium densocellulare complex, did vary seasonally. In particular, the volume of this brain region was significantly greater in November than in any of the other months that were sampled. As the authors note, the intermediate and medial mesopallium (IMM; previously known as intermediate and medial hyperstriatum ventrale or IMHV, see Horn, 1985), has been shown to be (part of) the neural substrate for imprinting memory (Horn, 1998, 2000, 2004). In addition, as discussed earlier, the caudomedial part of the mesopallium has been found to be involved in song memory in zebra finches (Bolhuis et al., 2000, 2001; Terpstra et al., 2004, 2006). Such seasonal variation, in food-storing birds, in the size of a brain region that has been shown to be involved in memory in two other avian paradigms should surely interest neuroecologists! Furthermore, Hoshooley and Sherry (2004) note that Lefebvre and colleagues (e.g., Lefebvre et al., 2002; Timmermans et al., 2000; Lefebvre and Bolhuis, 2003) have shown that the volume of the mesopallium/nidopallium complex correlates significantly with feeding innovations (see section on modularity, above).

Pravosudov and Omanska (2005) suggested an alternative explanation of Barnea and Nottebohm's (1994) results, namely that the seasonal trend in neurogenesis suggested by these authors "is not typical for food-caching animals" (p. 39). They add that "In any case, the significance of seasonal variation in neurogenesis in food-caching birds remains unclear unless we can confirm that memory capacity changes along with seasonal caching rates." This is reminiscent of Bolhuis and Macphail's (2001) earlier statement that "there is no evidence from any source to suggest that there are seasonal variations in memory capacity of storing birds" (p. 428). In addition, it is not clear why food-storing birds would need increased survival of new hippocampal neurons in October, when they store food, but not in the Winter, when they rely on memory to retrieve the previously stored food items.

Recent evidence from seasonally food-storing mammals is consistent with the suggestion that there is no correlation between food storing and volume or neuron number of, or neurogenesis in the hippocampus (Lavenex et al., 2000a, b). Lavenex et al. (2000a) reported that there was no significant seasonal variation in neuron number or neurogenesis in the dentate gyrus of the hippocampus (the brain region where there is continuous neurogenesis in adult mammals) of adult eastern grey squirrels (*Sciurus carolinensis*). Brain measurements were conducted in October (when the animals store food), January, and June. In a subsequent analysis of the same animals, Lavenex *et al.* (2000b) did not find significant seasonal variation in the relative volume (relative to the rest of the brain) of the hippocampus or of three subdivisions of the hippocampus, the dentate gyrus, CA1 and CA3. In addition, these authors did not find seasonal variation in either neuron number or neuron density in the hippocampus or its subregions. Interestingly, there was significant seasonal variation in the volume of the whole brain (excluding the cerebellum), with brains being largest in October and smallest in June.

Two studies have investigated the effects of artificially induced seasonality – by means of manipulation photoperiod – on hippocampus volume in black-capped chickadees (Krebs *et al.*, 1995; MacDougall-Shackleton *et al.*, 2003). This can be seen as a more controlled way of analyzing possible seasonal effects, as photoperiod is the only variable that is manipulated and e.g., climatic changes are excluded. Krebs *et al.* (1995) switched chickadees from long days to short days. The authors successfully mimicked a seasonal switch to Autumn, as the birds significantly increased their food storing rate. There was no change in the volume of the hippocampus. Recently, MacDougall-Shackleton *et al.* (2003) conducted a complementary experiment by switching chickadees from a short day schedule to a long day schedule (the "photostimulated" group). These authors included two control groups, one that was maintained on long days and one that was maintained on a short day schedule. The birds on the short day ("Autumn") schedule stored significantly more food than those on the long day schedule, and the birds in the photostimulated group stored significantly less food after they were switched to the long day schedule. Again, there was no correlation between photoperiod-induced food storing and volume of the hippocampus. Although photoperiod had a significant effect on storing behavior in both studies, it could be argued that under laboratory conditions the animals did not have sufficient opportunity to store food to be able to see any reflections of their behavior in hippocampal volume. Given this limitation, however, these results do not support the hypothesis of a relationship between food storing behavior and hippocampal volume.

Taken together, there is little evidence to suggest seasonal variation in the volume of the hippocampus, the number of neurons in the hippocampus or hippocampal neurogenesis in food-storing birds or mammals that correlates with seasonal variation in food storing. It is possible that in chickadees, there is seasonal variation in survival of newly generated hippocampal neurons (Barnea and Nottebohm, 1994;

Hoshooley and Sherry, 2004), that would be accompanied by seasonal variation in cell death, as total neuron numbers in the hippocampus do not vary seasonally. The fact that survival of newly generated neurons is greatest in October, when chickadees store food, is suggestive, but the significance of this phenomenon for learning and memory remains to be elucidated. Interestingly, a number of studies have shown that, although there is no seasonal variation in the volume of the hippocampus, there is significant variation in the volume of the whole brain (Lavenex *et al.*, 2000b), of the telencephalon (Smulders *et al.*, 1993; Smulders *et al.*, 1995, according to the reanalysis conducted by Lavenex *et al.*, 2000b) and of the mesopallium/hyperpallium densocellulare complex (Hoshooley and Sherry, 2004).

## THE NEUROECOLOGY DEBATE: WHERE DO WE STAND?

### Reflections of the debate in recent neuroecological publications

The neuroecology debate has led to some changes in the interpretation of recent relevant findings, and it is often acknowledged that the adaptationist approach to cognition is not without controversy. For instance, Shettleworth and Westwood (2002) stated that "Adaptation and modularity (…) are surprisingly controversial when it comes to learning and memory in humans or other species (…). The recent critical reviews of research on memory in food-storing birds and some other candidates for adaptive specializations of cognition by Macphail and Bolhuis (2001; Bolhuis and Macphail, 2001) are simply the newest manifestation of this controversy" (p. 240). At the same time, it is acknowledged that the results are not always consistent with an adaptationist interpretation. For instance, in their investigation of spatial and non-spatial learning in a food-storing (chickadees) and a non-storing species (dark eyed juncos), Shettleworth and Westwood (2002) state that "However, the evidence for the chickadees having a quantitative adaptive specialization in any aspect of spatial memory, as compared with the juncos, is very limited in the present data in that the chickadees did not uniformly outperform the juncos on spatial tests" (p. 238). They conclude that "Certainly the mismatch between the robustness of species differences in both behavior and brain anatomy and the fragility of species differences in spatial cognition pointed to by Bolhuis and Macphail (2001) poses a central problem for future

research in this particular corner of the adaptationist study of mind."
Pravosudov *et al.* (2002) concluded that "Our results therefore provide
further experimental evidence in support of the view that changes
in food-caching behavior and/or spatial memory performance do not
necessarily correlate with changes in hippocampal structure in fully
developed experienced food-caching animals (…)" (p. 145). In a recent
review, Shettleworth (2003) asked: "What accounts for these inconsist-
encies? One possibility is that the [adaptationist] hypothesis is too
naïvely simple. Or perhaps (…) it is framed in the wrong way." Brodin
and Lundborg (2003) go one step further when they state: "It is possible
that the adaptionistic approach used by neuroecologists may be wrong
and that a general-purpose approach (Macphail and Bolhuis, 2001) is
more correct" (p. 1562).

Indeed, the present review of recent studies lends little support
to the neuroecological hypothesis of a relationship between the morphology of – in my view, arbitrarily chosen – brain regions and particular cognitive mechanisms. In answer to Shettleworth's (1998) crucial
question, I maintain that function cannot explain mechanism. What
remains is the possibility that a neuroecological approach might provide clues for the study of the mechanisms of cognition and behavior.
I will evaluate this possibility in the final section of this chapter.

### Looking for clues

In my view, then, the only value of a neuroecological approach to the
study of the neural mechanisms of learning and memory would be to
provide clues for the investigation of the underlying mechanisms.
Dwyer and Clayton (2002) agree with this when they say that "(…)
Bolhuis and Macphail are right to emphasize that functional considerations can only provide clues for the investigation of the mechanisms
of behaviour (neural and otherwise)" (p. 110). Similarly, Healy *et al.*
(2005) suggested that "[neuroecologists] take functional considerations
as providing *clues* as to what the structure of the causal mechanisms
*might be*" (p. 20; my italics). This is akin to what Shettleworth (1998) has
called the "shallow synthesis" where "researchers working within one
program are forced to look outside it for answers to some of their
questions, but the core questions of the program are not at issue"
(p. 576). I maintain that researchers of animal cognition are not "forced"
to turn to the "ecological program," but they may derive useful clues
from it. As we have seen, such clues can be misleading and they have
been shown to be wrong on several occasions. This is not surprising, as

neuroecological clues are not intrinsically better than other clues. The philosopher David Hume identified the naturalistic fallacy, whereby certain moral principles were given legitimacy because they were thought to have been derived from the natural state of things. It would be equally fallacious to attribute a kind of scientific superiority to clues derived from evolutionary or functional considerations. As Shettleworth (2003) put it:

> Now, it is absolutely correct that knowing, for example, that black-capped chickadees are under intense selection for successful retrieval of stored food does not tell us exactly how they actually manage to retrieve any food. The supposition based on natural history, experimental data, and modeling that food-storing is important to their winter survival and subsequent reproduction tells us only that they must have cognitive and neural mechanisms allowing them to retrieve their scatter-hoarded food. It certainly does not tell us whether they use memory, or their hippocampus, let alone exactly how these work, never mind how – if at all – memory or hippocampi differ among chickadee populations or species. (p. 115)

Thus, a neuroecological approach by itself is clearly insufficient. How do we move from here? I agree with many neuroecologists that there are advantages in using "naturalistic" tasks when studying learning and memory. So, rather than using pigeons in skinnerboxes or rats in shuttleboxes, it is sensible to study naturally occurring behavior such as occurs in imprinting and song learning in birds or olfactory learning in rodents. In other words, animals' performance in cognitive tasks should be studied in situations where they can display their abilities optimally. As we have seen, because of the many sources of variation in the field, it is often inevitable that animals are taken to the laboratory to enable controlled experimental investigation. On the other hand, captivity may cause artifacts that can result in spurious findings. For instance, opportunities for food storing are relatively limited in a laboratory situation. Also, animals may behave differently in the laboratory compared to the field, and captivity may induce stress that can affect the results.

Clearly, there is endless variation in animal behavior. Songbirds are better at singing than non-songbirds, food storers are better at storing food than non-storers. Such differences do not necessarily imply that these species also differ in their cognitive capacities. We will not know unless we test adaptationist hypotheses empirically. I maintain that the evidence to date does not support the idea of major qualitative differences in cognitive abilities, just as Darwin suggested. It may well be that differences in performance in cognitive tasks that have been found are due to evolutionary specialization in modular input systems (Lefebvre and Bolhuis, 2003), along the lines of Fodor's (1983) original suggestions. This is not to say that functional and evolutionary clues should be disregarded altogether. As the maestro put it:

> In spite of the fact that we shall never be able to prove directly which contributions selection has made in the past, and that therefore any conclusion about the way interaction with the environment has moulded present-day species must remain tentative and, as such, different from conclusions drawn in the field of Physiology, Ontogeny and Survival Value, the ethologist feels that this is no reason to dismiss evolutionary study as just speculation. (Tinbergen, 1963, p. 429)

Indeed, we need not dismiss functional and evolutionary clues, but when studying the mechanisms of behavior, neuroecological clues should be treated with appropriate caution, and should be tested in controlled conventional experimental paradigms. Clues are just that; they are not explanations.

### SUMMARY AND CONCLUSIONS

The four questions that Niko Tinbergen identified for behavioral biology – concerning evolution, function, development, and causation – are all important and should be studied in their own right. Recently, there has been a debate as to whether these four questions should be investigated separately or they should be integrated. Integration of the four questions has been attempted in novel research disciplines such as cognitive ecology, evolutionary psychology, and neuroecology. Euan Macphail and I have criticized these integrative approaches, suggesting that they are fundamentally flawed, as they confound function and mechanism. Investigating the function or evolutionary history of a behavior or cognitive system is important and entirely legitimate. However, such investigations cannot provide us with answers to questions about the mechanisms underlying behavior or cognition. At most, functional or evolutionary considerations can provide clues that may be useful for a causal analysis of the underlying mechanisms. However, these clues can be misleading and are often wrong, as is illustrated with examples from song learning and food storing in birds. After summarizing the main issues in the neuroecology debate, I discuss some misunderstandings that were apparent in the replies to our critique, as well as some recent relevant data. Recent results do not support the neuroecological approach. Finally, I suggest that the way forward is a cautious and critical use of functional and evolutionary clues in the study of the mechanisms of behavior.

### ACKNOWLEDGMENTS

I am grateful to Euan Macphail for thinking and writing with me on evolution, cognition, and the brain. I thank him and the following for stimulating discussions: Robert Biegler, Eliot Brenowitz, Selvino de Kort, Luc-Alain Giraldeau, Rob Hampton, Marc Hauser, Jerry Hogan, Kevin Laland, Louis Lefebvre, Herman Philipse, Simon Reader, David Sherry, and Sara Shettleworth. Some of these authors disagreed with me, sometimes violently, but the discussion has (almost) always been fair and I hope that it will continue that way. I thank Jerry Hogan, David Sherry, and Simon Verhulst for their comments on an earlier version of the manuscript.

# References

Abrams, P. (2001). Adaptationism, optimality models and tests of adaptive scenarios. In *Adaptationism and Optimality*, ed. S.H. Orzack and E. Sober. Cambridge: Cambridge University Press, pp. 273–302.

Ainsworth, M.D.S., Blehar, M.C., Waters, E., and Wall, S. (1978). *Patterns of Attachment: Assessed in the Strange Situation and at Home*. Hillsdale, NJ: Erlbaum.

Airey, D.C. and DeVoogd, T.J. (2000). Greater song complexity is associated with augmented song system in zebra finches. *NeuroReport*, **11**, 2339–2344.

Airey, D.C., Buchanan, K.L., Szekely, T., Catchpole, C.K., and DeVoogd, T.J. (2000). Song, sexual selection, and song control nucleus (HVC) in the brains of European sedge warblers. *J. Neurobiol.*, **44**, 1–6.

Alcock, J.S. and Sherman, P.W. (1994). The utility of the proximate–ultimate dichotomy in ethology. *Ethology*, **96**, 58–62.

Altman, J. and Das, G.D. (1965). Autoradiographic and histological evidence of postnatal hippocampal neurogenesis in the rat. *J. Comp. Neurol.*, **124**, 319–355.

Amundson, R. (1996). Historical development of the concept of adaptation. In *Adaptation*, ed. M.R. Rose and G.V. Lauder. San Diego, CA: Academic Press, pp. 11–53.

Andersson, M. (1994). *Sexual Selection*. Princeton, NJ: Princeton University Press.

Arnold, A.P. (2002). Concepts of genetic and hormonal induction of vertebrate sexual differentiation in the twentieth century, with special reference to the brain. In *Hormones, Brain, and Behavior*, ed. D.W. Pfaff, A. Arnold, A. Etgen, S. Fahrbach, R. Moss, and R. Rubin, Volume 4. New York: Academic Press, pp. 105–136.

Aronson, L.R. (1981). Evolution of telencephalic function in lower vertebrates. In *Evolution of Telencephalic Function in Lower Vertebrates*, ed. P.R. Laming. Cambridge: Cambridge University Press, pp. 33–58.

Atz, J.W. (1970). The application of the idea of homology to behavior. In *Development and Evolution of Behavior: Essays in Honor of T.C. Schneirla*, ed. L.R. Aronson; E. Tobach, D.S. Lehrman, and J.S. Rosenblatt. San Francisco: W.H. Freeman, pp. 53–74.

Autumn, K., Ryan, M.J., and Wake, D.B. (2002). Integrating historical and mechanistic biology enhances the study of adaptation. *Q. Rev. Biol.*, **77**, 383–408.

Baerends, G.P. (1975). An evaluation of the conflict hypothesis as an explanatory principle for the evolution of displays. In *Function and Evolution in Behaviour*, ed. G.P. Baerends, C. Beer, and A. Manning. London: Oxford University Press, pp. 187–227.

Baerends, G.P. (1976). The functional organization of behaviour. *Anim. Behav.*, **24**, 726–738.

Baerends, G.P. and Drent, R.H., (eds.) (1982). The herring gull and its egg. Part II. *Behaviour*, **82**, 1–416.

Baerends, G.P. and Kruijt, J.P. (1973). Stimulus selection. In *Constraints on Learning*, ed. R.A. Hinde and J. Stevenson-Hinde. London: Academic Press, pp. 23–49.

Baker, J.R. (1938). The evolution of breeding seasons. In *Evolution: Essays on Aspects of Evolutionary Biology*, ed. G.R. de Beer. Oxford, UK: Oxford University Press, pp. 161–171.

Bakker, T.C.M. (1986). Aggressiveness in sticklebacks (*Gasterosteus aculeatus* L.).: a behaviour–genetic study. *Behaviour*, **98**, 1–144.

Balda, R.P., Kamil, A.C., Bednekoff, P.A., and Hile, A.G. (1997). Species differences in spatial memory performance on a three-dimensional task. *Ethology*, **103**, 47–55.

Ball, G.F., Riters, L.V., and Balthazart, J. (2002). Neuroendocrinology of song behavior and avian brain plasticity: multiple sites of action of sex steroid hormones. *Front. Neuroendocrin.*, **23**, 137–178.

Baptista, L.F. and Gaunt, S.L.L. (1997). Social interaction and vocal development in birds. In *Social Influences on Vocal Development*, ed. C.T. Snowdon and M. Hausberger. Cambridge, UK: Cambridge University Press, pp. 23–40.

Baptista, L.F. and Petrinovich, L. (1984). Social interaction, sensitive phases and the song template hypothesis in the white-crowned sparrow. *Anim. Behav.*, **34**, 1359–1371.

Barnard, C. (2004). *Animal Behaviour: Mechanism, Development, Function and Evolution.* Harlow, UK: Pearson Education Ltd/Prentice Hall.

Barnea, A. and Nottebohm, F. (1994). Seasonal recruitment of hippocampal neurons in adult free-ranging black-capped chickadees. *Proc. Natl Acad. Sci. USA*, **91**, 11217–11221.

Barnea, A. and Nottebohm, F. (1996). Recruitment and replacement of hippocampal neurons in young and adult chickadees: an addition to the theory of hippocampal learning. *Proc. Natl Acad. Sci. USA*, **93**, 714–718.

Basil, J.A., Kamil, A.C., Balda, R.P., and Fite, K.V. (1996). Differences in hippocampal volume among food storing corvids. *Brain Behav. Evol.*, **47**, 156–164.

Bateson, P. (1982). Preferences for cousins in Japanese quail. *Nature*, **295**, 236–237.

Bateson, P. (1987). Imprinting as a process of competitive exclusion. In *Imprinting and Cortical Plasticity*, ed. J.P. Rauschecker and P. Marler. New York: Wiley, pp. 151–168.

Bateson, P. (2003). The promise of behavioural biology. *Anim. Behav.*, **65**, 1–17.

Bateson, P.P.G. (1964). Effect of similarity between rearing and testing conditions on chicks' following and avoidance responses. *J. Comp. Physiol. Psychol.*, **57**, 100–103.

Bateson, P.P.G. (1966). The characteristics and context of imprinting. *Biol. Rev.*, **41**, 177–220.

Bateson, P. and Hinde, R.A. (1987). Developmental changes in sensitivity to experience. In *Sensitive Periods in Development*, ed. M.H. Bornstein. Hillsdale, NJ: Erlbaum, pp. 19–34.

Bateson, P. and Horn, G. (1994). Imprinting and recognition memory: a neural net model. *Anim. Behav.*, **48**, 695–715.

Bateson, P. and Martin, P. (2002). *Design for a Life. How Behaviour Develops.* London: Jonathan Cape.

Bateson, P.P.G. and Klopfer, P.H. (eds.) (1989). *Whither Ethology? Perspectives in Ethology*, vol. 8. New York: Plenum Press.

Beach, F.A. (1948). *Hormones and Behavior*. New York: Hoeber, Harper and Brothers.

Beach, F.A. and Jaynes, J. (1954). Effects of early experience upon the behavior of animals. *Psych. Bull.*, **51**, 239–263.

Bennett, A.T.D., Cuthill, I.C., Partridge, J.C., and Lunau, K. (1997). Ultraviolet plumage colors predict mate preferences in starlings. *Proc. Natl Acad. Sci., USA*, **94**, 8618–8621.

Bern, H.A. (1990). The "new" endocrinology: its scope and its impact. *Amer. Zool.*, **30**, 877–885.

Bernard, D.J., Eens, M., and Ball, G.F. (1996). Age and behavior-related variation in the volume of song control nuclei in male European starlings. *J. Neurobiol.*, **30**, 329–339.

Berridge, K.C. (1994). The development of action patterns. In *Causal Mechanisms of Behavioural Development*, ed. J.A. Hogan and J.J. Bolhuis. Cambridge, UK: Cambridge University Press, pp. 147–180.

Berridge, K.C., Fentress, J.C., and Parr, H. (1987). Natural syntax rules control action sequence of rats. *Behav. Brain Res.*, **23**, 59–68.

Bischof, H.-J. (1979). A model of imprinting evolved from neurophysiological concepts. *Z. Tierpsychol.*, **51**, 126–139.

Bischof, H.-J. (1994). Sexual imprinting as a two-stage process. In *Causal Mechanisms of Behavioural Development*, ed. J.A. Hogan and J.J. Bolhuis. Cambridge, UK: Cambridge University Press, pp. 82–97.

Bischof, H.-J. (2003). Neural mechanisms of sexual imprinting. *Anim. Biol.*, **53**, 89–112.

Blass, E.M. (1999). The ontogeny of human infant fact recognition: orogustatory, visual, and social influences. In *Early Social Cognition*, ed. P. Rochat. Mahway, NJ: Erlbaum, pp. 35–65.

Blokhuis, H.J. (1989). *The development and causation of feather pecking in the domestic fowl*. Ph.D. thesis, Agricultural University, Wageningen, the Netherlands.

Blokhuis, H.J. and Arkes, J.G. (1984). Some observations on the development of feather-pecking in poultry. *Appl. Anim. Behav. Sci.*, **12**, 145–157.

Blomberg, S.P. and Garland, T.J. (2002). Tempo and mode in evolution: phylogenetic inertia, adaptation and comparative methods. *J. Evol. Biol.*, **15**, 899–910.

Bock, W.J. and von Wahlert, G. (1965). Adaptation and the form–function complex. *Evolution*, **19**, 269–299.

Boesch, C. and Boesch, H. (1984). Mental maps in wild chimpanzees: an analysis of hammer transports for nut cracking. *Primates*, **25**, 160–170.

Bolhuis, J.J. (1991). Mechanisms of avian imprinting: a review. *Biol. Rev.*, **66**, 303–345.

Bolhuis, J.J. (1996). Development of perceptual mechanisms in birds: predispositions and imprinting. In *Neuroethological Studies of Cognitive and Perceptual Processes*. ed. C.F. Moss and S.J. Shettleworth. Boulder, CO: Westview Press, pp. 158–184.

Bolhuis, J.J. (1999). The development of animal behaviour. From Lorenz to neural nets. *Naturwissenschaften*, **86**, 101–111.

Bolhuis, J.J. (ed.). (2000). *Brain, Perception, Memory. Advances in Cognitive Neuroscience*. Oxford: Oxford University Press.

Bolhuis, J.J. (2005a). Development of behavior. In *The Behavior of Animals. Mechanisms, Function, and Evolution*, ed. J.J. Bolhuis and L.-A. Giraldeau. Oxford: Blackwell Publishing, pp. 119–145.

Bolhuis, J.J. (2005b). Function and mechanism in neuroecology: looking for clues. *Anim. Biol.* **55**, 457–490. (Chapter 9 in this volume).

Bolhuis, J.J. (2008). Chasin' the trace: the neural substrate of bird song memory. In *The Neuroscience of Birdsong*, (ed. H.P. Zeigler and P. Marler. Cambridge, UK: Cambridge University Press).

Bolhuis, J.J. and Eda-Fujiwara, H. (2003). Bird brains and songs: neural mechanisms of birdsong perception and memory. *Anim. Biol.* **53**, 129–145.

Bolhuis, J.J. and Gahr, M. (2006). Neural mechanisms of birdsong memory. *Nature Rev. Neurosci.*, **7**, 347–357.

Bolhuis, J.J. and Giraldeau, L.-A. (2005). The study of animal behavior. In *The Behavior of Animals. Mechanisms, Function, and Evolution*, ed. J.J. Bolhuis and L.-A. Giraldeau. Oxford: Blackwell Publishing, pp. 1–9.

Bolhuis, J.J. and Hogan, J.A. (eds.) (1999). *The Development of Animal Behavior: A Reader*. Oxford, UK: Blackwell.

Bolhuis, J.J. and Honey, R.C. (1994). Within-event learning during filial imprinting. *J. Exp. Psychol.: Anim. Behav. Processes*, **20**, 240–248.

Bolhuis, J.J. and Honey, R.C. (1998). Imprinting, learning, and development: from behaviour to brain and back. *Trends Neurosci.*, **21**, 306–311.

Bolhuis, J.J. and Macphail, E.M. (2001). A critique of the neuroecology of learning and memory. *Trends Cogn. Sci.*, **5**, 426–433.

Bolhuis, J.J. and Macphail, E.M. (2002). Everything in neuroecology makes sense in the light of evolution. *Trends Cogn. Sci.*, **6**, 7–8.

Bolhuis, J.J., de Vos, G.J., and Kruijt, J.P. (1990). Filial imprinting and associative learning. *Q. J. Exp. Psychol.*, **42B**, 313–329.

Bolhuis, J.J., Zijlstra, G.G.O., Den Boer-Visser, A.M., and Van der Zee, E.A. (2000). Localized neuronal activation in the zebra finch brain is related to the strength of song learning, *Proc. Natl Acad. Sci. USA*, **97**, 2282–2285.

Bolhuis, J.J., Hetebrij, E., Den Boer-Visser, A.M., De Groot, J.H., and Zijlstra, G.G.O. (2001). Localized immediate early gene expression related to the strength of song learning in socially reared zebra finches, *Eur. J. Neurosci.*, **13**, 2165–2170.

Booth, D.A. (ed.) (1978). *Hunger Models*. London: Academic Press.

Borbély, A.A. (1982). A two-process model of sleep regulation. *Hum. Neurobiol.*, **1**, 195–204.

Borbély, A.A. and Achermann, P. (1999). Sleep homeostasis and models of sleep regulation. *J. Biol. Rhythms*, **14**, 557–568.

Bornstein, M.H. (ed) (1987). *Sensitive Periods in Development*. Hilldale, NJ: Erlbaum.

Bottjer, S.W. and Arnold, A.P. (1997). Developmental plasticity in neural circuits for a learned behavior. *Annu. Rev. Neurosci.*, **20**, 459–481.

Bottjer, S.W., Miesner, E.A., and Arnold, A.P. (1984). Forebrain lesions disrupt development but not maintenance of song in passerine birds. *Science*, **224**, 901–903.

Boul, K.E. and Ryan, M.J. (2004). Population variation of complex advertisement calls in *Physalaemus petersi* and comparative laryngeal morphology. *Copeia*, **3**, 624–631.

Bourne, G.R. (1993). Proximate costs and benefits of mate acquisition at leks of the frog *Ololygon rubra*. *Anim. Behav.*, **45**, 1051–1059.

Bowlby, J. (1969). *Attachment and Loss, Vol. 1. Attachment*. London: Hogarth Press.

Bowlby, J. (1991). Ethology and psychoanalysis. In *The Development and Integration of Behavior*, ed. P. Bateson. Cambridge, UK: Cambridge University Press, pp. 301–313.

Bowmaker, J.K. (1998). Evolution of colour vision in vertebrates. *Eye*, **12**, 541–547.

Brainard, M.S. and Doupe, A.J. (2000). Auditory feedback in learning and maintenance of vocal behaviour. *Nature Rev. Neurosci.*, **1**, 31–40.

Brainard, M.S. and Doupe, A.J. (2002). What songbirds teach us about learning. *Nature*, **417**, 351–358.

Brandon, R.N. (1990). *Adaptation and Environment*. Princeton, NJ: Princeton University Press.

Brenowitz, E.A. (1991). Altered perception of species-specific song by female birds after lesions of a forebrain nucleus. *Science*, **251**, 303–304.

Brenowitz, E.A. and Beecher, M.D. (2005). Song learning in birds: diversity and plasticity, opportunities and challenges. *Trends Neurosci.*, **28**, 127–132.

Brenowitz, E.A. and Kroodsma, D.E. (1996). The neuroethology of birdsong. In *Ecology and Evolution of Acoustic Communication in Birds*, ed. D.E. Kroodsma and E.H. Miller. London: Cornell University Press, pp. 285–304.

Brenowitz, E.A., Nalls, B., Wingfield, J., and Kroodsma, D. (1991). Seasonal changes in avian song nuclei without seasonal changes in song repertoire. *J. Neurosci.*, **11**, 1367–1374.

Brenowitz, E.A., Lent, K., and Kroodsma, D.E. (1995). Brain space for learned song develops independently of song learning. *J. Neurosci.* **15**, 6281–6286.

Brodin, A. (2005). Hippocampus volume does not correlate to food hoarding rates in the field in two closely related bird species, the black-capped chickadee *Poecile atricapillus* and the willow tit *Parus montanus*. *Auk*, **122**, 819–828.

Brodin, A. and Lundborg, K. (2003). Is hippocampus volume affected by specialisation for food hoarding in birds? *Proc. Roy. Soc. Lond. B*, **270**, 1555–1563.

Brooks, D.R. and McLennan, D.A. (1991). *Phylogeny, Ecology, and Behavior*. Chicago: University of Chicago Press.

Brooks, D.R. and McLennan, D.A. (2002). *The Nature of Diversity*. Chicago: University of Chicago Press.

Brown, M.W. (2000). Neuronal correlates of recognition memory. In *Brain, Perception, Memory. Advances in Cognitive Neuroscience*, ed. J.J. Bolhuis. Oxford: Oxford University Press, pp. 185–208.

Bucher, T.L., Ryan, M.J., and Bartholomew, G.W. (1982). Oxygen consumption during resting, calling and nest building in the frog *Physalaemus pustulosus*. *Physiological Zoology*, **55**, 10–22.

Buckley, M.J. and Gaffan, D. (2000). The hippocampus, perirhinal cortex and memory in the monkey. In *Brain, Perception, Memory. Advances in Cognitive Neuroscience*, ed. J.J. Bolhuis. Oxford, UK: Oxford University Press, pp. 279–298.

Buller, D.J. (2005). *Adapting Minds. Evolutionary Psychology and the Persistent Quest for Human Nature*. Cambridge, MA and London: MIT Press.

Burghardt, G.M. and Gittleman, J.G. (1990). Comparative and phylogenetic analyses: new wine, old bottles. In *Interpretation and Explanation in the Study of Behavior: Vol. 2. Comparative Perspectives*, ed. M. Bekoff and D. Jamieson. Boulder, CO: Westview Press, pp. 192–225.

Burkhardt, R.W., Jr. (2005). *Patterns of Behavior. Konrad Lorenz, Niko Tinbergen and the Founding of Ethology*. Chicago and London: University of Chicago Press.

Byrne, R.W. (2007). Animal cognition: bring me my spear. *Current Biol.*, **17**, R164–R165.

Cannatella, D.C. and Duellman, W.E. (1984). Leptodactylid frogs of the *Physalaemus pustulosus* group. *Copeia* **4**, 902–921.

Cannatella, D.C., Hillis, D.M., Chippinendale, P., Weigt, L., Rand, A.S., and Ryan, M.J. (1998). Phylogeny of frogs of the *Physalaemus pustulosus* species group, with an examination of data incongruence. *Syst. Biol.*, **47**, 311–335.

Capranica, R.R. (1977). Auditory processing in anurans. *Fed. Proc.*, **37**, 2324–2328.

Carew, T.J. (2000). *Behavioral Neurobiology: The Cellular Organization of Natural Behavior*. Sunderland, MA: Sinauer.

Carr, C.E. and Konishi, M. (1990). A circuit for detection of interaural time differences in the brain stem of the barn owl. *J. Neurosci.*, **10**, 3227–3246.

Catchpole, C.K. and Slater, P.J.B. (1995). *Bird Song: Biological Themes and Variations*. Cambridge, UK: Cambridge University Press.

Cheney, D.L. and Seyfarth, R.M. (1990). *How Monkeys See the World: Inside the Mind of Another Species*. Chicago: University of Chicago Press.

Chittka, L. (1996). Does bee color vision predate the evolution of flower color? *Naturwissenschaften*, **83**, 136–138.

Chittka, L. and Briscoe, A. (2001). Why sensory ecology needs to become more evolutionary – insect color vision as a case in point. In *Ecology of Sensing*, ed. F.G. Barth and A. Schmid. Berlin: Springer-Verlag, pp. 19–37.

Chittka, L. and Menzel, R. (1992). The evolutionary adaptation of flower colours and the insect pollinators' colour vision. *J. Comp.Physiol. A*, **171**, 171–181.

Chomsky, N. (1965). *Aspects of the Theory of Syntax*. Cambridge, MA: MIT Press.

Chow, A. and Hogan, J.A. (2005). The development of feather pecking in Burmese red junglefowl: the influence of early experience with exploratory-rich environments. *Appl. Anim. Behav. Sci.*, **93**, 283–294.

Clark, M.M. and Galef, B.G. (1995). Prenatal influences on reproductive life history strategies. *Trends Ecol. Evol.*, **10**, 151–153.

Clark, M.M., Crews, D., and Galef, B.G. (1991a). Concentrations of sex steroid hormones in pregnant and fetal mongolian gerbils. *Physiol. Behav.*, **49**, 239–243.

Clark, M.M., Galef, B.G., and vom Saal, F.S. (1991b). Nonrandom sex composition of gerbil mouse and hamster litters before and after bird. *Devel. Psychobiol.*, **24**, 81–90.

Clayton, D.F. (2000). The neural basis of avian song learning and perception. In *Brain, Perception, Memory. Advances in Cognitive Neuroscience*, ed. J.J. Bolhuis. Oxford University Press, Oxford: pp. 113–125.

Clayton, N.S. (1994). The influence of social interactions on the development of song and sexual preferences in birds. In *Causal Mechanisms of Behavioural Development*, ed. J.A. Hogan and J.J. Bolhuis. Cambridge, UK: Cambridge University Press, pp. 98–115.

Clayton, N.S. and Dickinson, A. (1998). Episodic-like memory during cache recovery by scrub jays. *Nature*, **395**, 272–274.

Clayton, N.S. and Krebs, J.R. (1995a). Memory in food-storing birds: from behaviour to brain. *Curr. Opin. Neurobiol.* **5**, 149–154.

Clayton, N.S. and Krebs J.R. (1995b). Lateralization in memory and the avian hippocampus in food-storing birds. In *Behavioural Brain Research in Naturalistic and Semi-naturalistic Settings*, ed. E. Alleva, A. Fasolo, H.P. Lipp, L. Nadel, and L. Ricceri. Dordrecht: Kluwer Academic Publishers, pp. 139–157.

Clayton, N.S., Bussey, T.J., and Dickinson, A. (2003). Can animals recall the past and plan for the future? *Nature Rev. Neurosci.*, **4**, 685–691.

Clayton, N.W., Salwiczek, L.H., and Dickinson, A. (2007). Episodic memory. *Current Biol.*, **17**, R189–R191.

Cleveland, A., Rocca, A.M., Wendt, E.L., and Westergaard, G.C. (2004). Transport of tools to food sites in tufted capuchin monkeys (*Cebus apella*). *Anim. Cogn.*, **7**, 193–198.

Clotfelter, E.D., O'Neal, D.M., Gaudioso, J.M. *et al.* (2004). Consequences of elevating plasma testosterone in females of a socially monogamous songbird: evidence of constraints on male evolution? *Horm. Behav.*, **46**, 171–178.

Colombo, M. and Broadbent, N. (2000). Is the avian hippocampus a functional homologue of the mammalian hippocampus? *Neurosci. Biobehav. Rev.*, **24**, 465–484.

Colonnese, M.T., Stallman, E.L., and Berridge, K.C. (1996). Ontogeny of action syntax in altricial and precocial rodents: grooming sequences of rat and guinea pig pups. *Behaviour*, **133**, 1165–1195.

Coomber, P., Crews, D., and Gonzalez-Lima, F. (1997). Independent effects of incubation temperature and gonadal sex on the volume and metabolic capacity of brain nuclei in the leopard gecko, *Eublepharis macularius*, a lizard with temperature-dependent sex determination. *J. Comp.Neurol.*, **380**, 409–421.

Cosmides, L. and Tooby, J. (1994). Origins of domain specificity: the evolution of functional organization. In *Mapping the Mind*, L.A. Hirschfeld and S.A. Gelman. Cambridge, UK: Cambridge University Press, pp. 85–116.

Cosmides, L. and Tooby, J. (1995). From function to structure: the role of evolutionary biology and computational theories in cognitive neuroscience. In *The Cognitive Neurosciences*, ed. M. Gazzaniga. Cambridge, MA, MIT Press, pp. 1199–1210.

Crews, D. (1992). 2-DG and neuroethology: Metabolic mapping of brain activity during sexual and aggressive species typical behaviors. In *Advances in Metabolic Mapping Techniques for Brain Imaging of Behavioral and Learning Functions*, ed. F. Gonzalez-Lima, T. Finkenstaedt, and H. Scheich. Dordrecht, the Netherlands: Kluwer Academic Publishers B.V., pp. 367–387.

Crews, D. (1993). The Organizational Concept and vertebrates without sex chromosomes. *Brain, Behav. Evol.*, **42**, 202–214.

Crews, D. (1996). Temperature-dependent sex determination: the interplay of steroid hormones and temperature. *Zool. Science*, **13**, 1–13.

Crews, D. (1999). Sexuality: the environmental organization of phenotypic plasticity. In *Reproduction in Context*, K. Wallen and J. Schneider. Cambridge: MIT Press, pp. 473–499.

Crews, D. (2002). Diversity and evolution of neuroendocrine mechanisms underlying reproductive behavior. In *Behavioral Endocrinology*, 2nd edn. ed. J. Becker, S.M. Breedlove, D. Crews, and M. McCarthy. Cambridge: MIT Press, pp. 223–288.

Crews, D. and Groothuis, T. (2005). Tinbergen's fourth question, ontogeny: sexual and individual differentiation. *Anim. Biol.*, **55**, 343–370 (Chapter 4 this volume).

Crews, D. and Silver, R. (1985). Reproductive physiology and behavior interactions in nonmammalian vertebrates. In *Handbook of Behavioral Neurobiology, Vol. 7: Reproduction*, ed. N.T. Adler, D. Pfaff, and R.W. Goy. New York: Plenum, pp. 101–182.

Crews, D., Wibbels, T., and Gutzke, W.H.N. (1989). Action of sex steroid hormones on temperature-induced sex determination in the snapping turtle (*Chelydra serpentina*). *Gen. Comp.Endocr.*, **76**, 159–166.

Crews, D., Sakata, J.T., and Rhen, T. (1998). Developmental effects on intersexual and intrasexual variation in growth and reproduction in a lizard with temperature-dependent sex determination. *Comp.Biochem. Physiol. (Part C).*, **119**, 229–241.

Crews, D., Fuller, T., Mirasol, E.G., Pfaff, D.W., and Ogawa, S. (2004). Postnatal environment affects behavior of adult transgenic mice. *Exper. Biol. Med.*, **229**, 935–939.

Crook, J.H. (1964). The evolution of social organization and visual communication in the weaver birds (Ploceinae). *Behaviour Suppl.*, **10**, 1–178.

Cullen, E. (1957). Adaptation in the Kittiwake to cliff-nesting. *Ibis*, **99**, 275–302.

Cuthill, I.C. (2002). Review of Barth, F.G. and Schmid, A. 2001: *Ecology of Sensing.* Springer-Verlag, Berlin. *Ethology*, **108**, 566.

Cuthill, I.C. (2005). The study of function in behavioural ecology. *Anim. Biol.* **55**, 399–417. (Chapter 6 this volume.)

Cuthill, I.C., Partridge, J.C., Bennett, A.T.D., Church, S.C., Hart, N.S., and Hunt, S. (2000). Ultraviolet vision in birds. *Adv. Stud. Behav.*, **29**, 159–214.

Czeisler, C.A., Dijk, D-J., and Duffy, J.F. (1994). Entrained phase of the circadian pacemaker serves to stabilize alertness and performance throughout the habitual waking day. In *Sleep Onset: Normal and Abnormal Processes*, ed. R.D. Ogilvie and J.R. Harsh. Washington, DC: Amer. Psychol. Assn., pp. 89–110.

Daan, S., Beersma, D.G.M., and Borbély, A.A. (1984). Timing of human sleep: recovery process gated by a circadian pacemaker. *Am. J. Physiol.*, **246**, R161–R183.

Daly, M. and Wilson, M. (1978). *Sex, Evolution, and Behavior*. North Scituate, MA: Duxbury Press.

Daly, M. and Wilson, M. (2005). Human behavior as animal behavior. In *The Behavior of Animals. Mechanisms, Function, and Evolution*, ed. J.J. Bolhuis and L.-A. Giraldeau. Oxford: Blackwell Publishing, pp. 393–408.

Daly, M. and Wilson, M.I. (1999). Human evolutionary psychology and animal behaviour. *Anim. Behav.*, **57**. 509–519.

Darwin, C. (1871). *The Descent of Man and Selection in Relation to Sex*. London: Murray.

Davies, N.B. (1992). *Dunnock Behaviour and Evolution*. Oxford: Oxford University Press.

Davies, N.B. and Halliday, T.R. (1978). Deep croaks and fighting assessment in toads *Bufo bufo. Nature*, **274**, 683–685.

Dawkins, M.S. (1986). *Unravelling Animal behaviour*. Harlow, UK: Longman.

Dawkins, M.S. (1989). The future of ethology: how many legs are we standing on? In *Whither Ethology? Perspectives in Ethology*, vol. 8. ed. P.P.G. Bateson and P.H. Klopfer. New York: Plenum Press, pp. 47–54.

Dawkins, M.S. (1990). From an animal's point of view – motivation, fitness, and animal-welfare. *Behav. Brain Sci.*, **13**, 1–61.

Dawkins, M.S. (2006). A user's guide to animal welfare science. *Trends Ecol. Evol.*, **21**, 77–82.

Dawkins, M.S. and Guilford, T. (1995). An exaggerated preference for simple neural network models of signal evolution? *Proc. Roy. Soc. Lond. B*, **261**, 357–360.

Dawkins, R. (1986). *The Blind Watchmaker*. Harlow, UK: Longman.

Dawkins, R. (1996). *Climbing Mount Improbable*. London: Viking.

De Kort, S., Dickinson, A., and Clayton, N.S. (2005). Retrospective cognition by food-caching western scrub jays. *Learning and Motivation*, **36**, 159–176.

Dennett, D.C. (1987). *The Intentional Stance*. Cambridge, MA: MIT Press.

Dennett, D.C. (1995). *Darwin's Dangerous Idea. Evolution and the Meaning of Life*. London: Penguin Books Ltd.

DeQueiroz, A. and Wimberger, P.H. (1993). The usefulness of behavior for phylogeny estimation: levels of homoplasy in behavioral and morphological characters. *Evolution*, **47**, 46–60.

DeVoogd, T.J. (1994). The neural basis for the acquisition and production of bird song. In *Causal Mechanisms of Behavioural Development*, ed. J.A. Hogan and J.J. Bolhuis. Cambridge: Cambridge University Press, pp. 49–81.

DeVoogd, T.J. and Székely, T. (1998). Causes of avian song: using neurobiology to integrate proximate and ultimate levels of analysis. In *Animal Cognition in Nature*, ed. R.P. Balda, I.M. Pepperberg, and A.C. Kamil. San Diego: Academic Press, pp. 337–380.

DeVoogd, T.J., Krebs, J.R. Healy, S.D., and Purvis, A. (1993). Relations between song repertoire size and the volume of brain nuclei related to song: comparative evolutionary analyses amongst oscine birds. *Proc. Roy. Soc. Lond. B*, **254**, 75–82.

Dewsbury, D.A. (1992). On the problems studied in ethology, comparative psychology, and animal behavior. *Ethology*, **92**, 89–107.

Dewsbury, D.A. (1999). The proximate and the ultimate: past, present, and future. *Behav. Proc.*, **46**, 189–199.

Dijk, D-J. and Lockley, S.W. (2002). Invited review: integration of human sleep–wake regulation and circadian rhythmicity. *J. Appl. Physiol.*, **92**, 852–862.

Dijk, D-J., Duffy, J.F., and Czeisler, C.A. (1992). Circadian and sleep/wake dependent aspects of subjective alertness and cognitive performance. *J. Sleep Res.*, **1**, 112–117.

Dingemanse, N.J. and Réale, D. (2005). Natural selection and animal personality. *Behaviour*, **142**, 1159–1184.

Dingemanse, N.J., Wright, J., Kazem, A.J.N., Thomas, D.K., Hickling, R., and Dawnay, N. (2007). Behavioural syndromes differ predictably between 12 populations of three-spined stickleback. *J. Anim. Ecol.*, **76**, 1128–1138.

Dobzhansky, T. (1973). Nothing in biology makes sense except in the light of evolution. *Am. Biol. Teach.*, **35**, 125–129.

Doupe, A.J. and Kuhl, P.K. (1999). Birdsong and human speech: common themes and mechanisms. *Ann. Rev. Neurosc.*, **22**, 567–631.

Duchaine, B., Cosmides, L., and Tooby, J. (2001). Evolutionary psychology and the brain. *Curr. Opin. Neurobiol.*, **11**, 225–230.

Duncan. I.J.H. (2002). Poultry welfare: science or subjectivity? *Brit. Poultry Sci.*, **43**, 643–652.

Dwyer, D.M. and Clayton, N.S. (2002). A reply to the defenders of the faith. *Trends Cogn. Sci.*, **6**, 109–111.

Eda-Fujiwara, H., Satoh, R., Bolhuis, J.J., and Kimura, T. (2003). Neuronal activation in female budgerigars is localized and related to male song complexity. *Eur. J. Neurosci.*, **17**, 149–154.

Eimas, P.D., Miller, J.L., and Jusczyk, P.W. (1987). On infant speech perception and the acquisition of language. In *Categorical Perception*, ed. S. Harnad. Cambridge, UK: Cambridge University Press, pp. 161–195.

Eimas, P.D., Siqueland, P., Jusczyk, P., and Vigorito, J. (1971). Speech perception in infants. *Science*, **171**, 303–306.

Eising, C.M. and Groothuis, T.G.G. (2003). Yolk androgens and begging behaviour in black-headed gull chicks: an experimental field study. *Anim. Behav.*, **66**, 1027–1034.

Eising, C.M., Eikenaar, C., Schwabl, H., and Groothuis, T.G.G. (2001). Maternal androgens in black-headed gull eggs: consequences for chick development. *Proc. R. Soc. Lond. B*, **268**, 839–846

Eising, C.M., Mueller, W., and Groothuis, T.G.G. (2006). Avian mothers produce different phenotypes by hormone deposition in their eggs. *Biol. Lett.* **2**, 20–22.

Emlen, J.M. (1966). The role of time and energy in food preference. *Am. Nat.*, **100**, 611–617.

Endler, J.A. and Basolo, A.L. (1998). Sensory ecology, receiver biases and sexual selection. *Trends Ecol. Evol.*, **13**, 415–420.

Enquist, M. and Arak, A. (1993). Selection of exaggerated male traits by female aesthetic senses. *Nature*, **361**, 446–448.

Enquist, M. and Arak, A. (1998). Neural representation and the evolution of signal form. In *Neural Representation and the Evolution of Signal Form*, ed. R. Dukas. Chicago: University of Chicago Press, pp. 21–87.

Enquist, M. and Ghirlanda, S. (2005). *Neural Networks and Animal Behavior*. Princeton, NJ: Princeton University Press.

Erichsen, J.T., Bingman, V.P., and Krebs, J.R. (1991). The distribution of neuropeptides in the dorsomedial telencephalon of the pigeon (*Columba livia*): a basis for regional subdivisions. *J. Comp.Neurol.*, **314**, 478–492.

Ewert, J.P. (1997). Neural correlates of key stimulus and releasing mechanism: a case study and two concepts. *Trends Neurosci.*, **20**, 332–339.

Fee, M.S., Mitra P.P., and Kleinfeld (1998). The role of nonlinear dynamics of the syrinx in the vocalizations of a songbird. *Nature*, **395**, 67–71.

Felsenstein, J. (1985). Phylogenies and the comparative method. *Am. Nat.*, **125**, 1–15.

Fentress, J.C. (1972). Development and patterning of movement sequences in inbred mice. In *The Biology of Behavior*, ed. J. Kiger. Corvallis, OR: Oregon State University Press, pp. 83–132.

Fentress, J.C. and Gadbois, S. (2001). The development of action sequences. In *Handbook of Behavioral Neurobiology, Vol. 13: Developmental Psychobiology*, ed. E.M. Blass. New York: Kluwer Academic/Plenum, pp. 393–431.

Fentress, J.C. and Stilwell, F.P. (1973). Grammar of a movement sequence in inbred mice. *Nature*, **224**, 52–53.

Fisher, R.A. (1930). *The Genetical Theory of Natural Selection*. Oxford: Clarendon Press.

Fleissner, G., Holtkamp-Rötzler, E., Hanzlik, M. *et al.* (2003). Ultrastructural analysis of a putative magnetoreceptor in the beak of homing pigeons. *J. Comp.Neurol.*, **458**, 350–360.

Fleming A.S. and Blass, E.M. (1994). Psychobiology of the early mother–young relationship. In *Causal Mechanisms of Behavioural Development*, ed. J.A. Hogan and J.J. Bolhuis. Cambridge, UK: Cambridge Univ. Press, pp. 212–241.

Flombaum, J.I, Santos, L.R., and Hauser, M.D. (2002). Neuroecology and psychological modularity. *Trends Cogn. Sci.*, **6**, 106–108.

Fodor, J.A. (1983). *The Modularity of Mind: An Essay on Faculty Psychology*. Cambridge, MA: MIT Press.

Fodor, J.A. (2000). *The Mind Doesn't Work That Way*. Cambridge, MA: MIT Press.

Francis, R.C. (1990). Causes, proximate and ultimate. *Biol. Phil.*, **5**, 401–415.

Freckleton, R.P., Harvey, P.H., and Pagel, M. (2002). Phylogenetic analysis and comparative data: a test and review of evidence. *Am. Nat.*, **160**, 712–726.

Gahr, M. (1997). How should brain nuclei be delineated? Consequences for developmental mechanisms and for correlations of area size, neuron numbers and functions of brain nuclei. *Trends Neurosci.*, **20**, 58–62.

Gahr, M., Leitner, S., Fusani, L., and Rybak, F. (2002). What is the adaptive role of neurogenesis in adult birds? In *Progress in Brain Research*, vol. 138, ed. M.A. Hofman, G.J. Boer, A.J.G.D. Holtmaat, E.J.W. van Someren, J. Verhaagen, and D.F. Swaab. Amsterdam: Elsevier Science, pp. 233–254.

Garamszegi, L.Z. and Eens, M. (2004a). Brain space for a learned task: strong intraspecific evidence for neural correlates of singing behavior in songbirds. *Brain Res. Rev.*, **44**, 187–193.

Garamszegi, L.Z. and Eens, M. (2004b). The evolution of hippocampus volume and brain size in relation to food hoarding in birds. *Ecol. Lett.*, **7**, 1216–1224.

Gerhardt, H.C. (1981). Mating call recognition in the barking treefrog (*Hyla gratiosa*): responses to synthetic calls and comparisons with the green treefrog (*Hyla cinerea*). *J. Comp.Physiol.*, **144**, 17–26.

Gerhardt, H.C. (2001). Acoustic communication in two groups of closely related treefrogs. In *Adv. Stud. Behav.*, Vol. 30, ed. P.J.B. Slater, J.S. Rosenblatt, C.T. Snowdon, and T.J. Roper. San Diego: Academic Press, pp. 99–167.

Gerhardt, H.C. and Schwartz, J.J. (2001). Auditory tunings and frequency preferences in anurans. In *Anuran Communication*, ed. M.J. Ryan. Washington DC: Smithsonian Institution Press, pp. 73–85.

Ghirlanda, S. (2002). Intensity generalization: physiology and modeling of a neglected topic. *J. Theor. Biol.*, **214**, 389–404.

Ghirlanda, S. and Enquist, M. (1998). Artificial neural networks as models of stimulus control. *Anim. Behav.*, **56**, 1383–1389.

Ghirlanda, S. and Enquist, M. (1999). The geometry of stimulus control. *Anim. Behav.*, **58**, 695–706.

Ghirlanda, S., Jansson, L., and Enquist, M. (2002). Chickens prefer beautiful humans. *Hum. Nature*, **13**, 383–389.

Gil, D., Graves, J., Hazon, N., and Wells, A. (1999). Male attractiveness and differential hormone investment in zebra finch eggs. *Science*, **286**, 126–128.

Gobes, S.M.H. and Bolhuis, J.J. (2007). Bird song memory: a neural dissociation between song recognition and production. *Curr. Biol.*, **17**, 789–793.

Goldin-Meadow, S. (1997). The resilience of language in humans. In *Social Influences on Vocal Development*, ed. C.T. Snowdon and M. Hausberger. Cambridge, UK: Cambridge University Press, pp. 293–311.

Goldman, S.A. and Nottebohm, F. (1983). Neuronal production, migration, and differentiation in a vocal control nucleus of the adult female canary brain. *Proc. Natl Acad. Sci. USA.*, **80**, 2390–2394.

Gonzalez-Lima, F. (1992). Brain imaging of auditory learning functions in rats: studies with fluorodeoxyglucose autoradiography and cytochrome oxidase histochemistry. In *Advances in Metabolic Mapping Techniques for Brain Imaging of Behavioral and Learning Functions*, ed. F. Gonzalez-Lima, T. Finkenstadt, and H. Scheich. Boston: Kluwer Academic Publ., pp. 39–109.

Goodale, M.A. and Westwood, D.A. (2004). An evolving view of duplex vision: separate but interacting cortical pathways for perception and action. *Curr. Opin. Neurobiol.*, **14**, 203–211.

Gottlieb, G. (1997). *Synthesizing Nature-Nurture: Prenatal Roots of Instinctive Behavior*. Mahway, NJ: Erlbaum.

Gould, S.J. and Lewontin, R.C. (1979). The spandrels of San Marco and the Panglossian paradigm: a critique of the adaptationist programme. *Proc. Roy. Soc. Lond. B*, **205**, 581–598.

Gould, S.J. and Vrba, E.S. (1982). Exaptation – a missing term in the science of form. *Paleobiology*, **8**, 4–15.

Gould-Beierle, K.L. (2000). A comparison of four corvid species in a working and reference memory task using a radial maze. *J. Comp.Psychol.*, **114**, 347–356.

Goyman, W, East, M.L., and Hofer, H. (2001). Androgens and the role of female "hyperaggresiveness" in spotted hyenas. *Horm. Behav.*, **39**, 83–92.

Grafen, A. (1984). Natural selection, kin selection and group selection. In *Behavioural Ecology. 2nd edn*, ed. J.R. Krebs and N.B. Davies. Oxford, UK: Blackwell Science Ltd., pp. 62–84.

Greenwalt, C.H. (1968). *Bird Song: Acoustics and Physiology*. Washington DC: Smithsonian Institution Press.

Griffin, D.R. (1976). *The Question of Animal Awareness*. New York: Rockefeller University Press.

Griffin, D.R. (1978). Prospects for a cognitive ethology. *Behav. Brain Sci.*, **1**, 527–538.

Griffin, D.R. (1992). *Animal Minds*. Chicago: University of Chicago Press.

Groothuis, T.G.G. (1992). The influence of social experience on the development and fixation of the form of displays in the black-headed gull. *Anim. Behav.*, **43**, 1–14.

Groothuis, T.G.G. (1993). Development of social displays: form development, form fixation and change in context. In *Advances in the Study of Behavior, Vol. 22*, ed. P.J.P. Slater, M. and Milinski, and J.S. Rosenblatt. New York: Academic Press, pp. 269–322.

Groothuis, T.G.G. (1994). The ontogeny of social displays: interplay between motor development, development of motivational systems and social experience. In *Causal Mechanisms of Behavioural Development*, ed. J.A. Hogan and J.J. Bolhuis. Cambridge, UK: Cambridge University Press, pp. 183–211.

Groothuis T. and van Mulekom, L. (1991). The influence of social experience on the ontogenetic change in the relation between aggression, fear and display behaviour in black-headed gulls. *Anim. Behav.*, **42**, 873–881.

Groothuis, T.G. and Schwabl, H. (2002). Determinants of within and among clutch variation in levels of maternal hormones in Black-headed gull eggs. *Funct. Ecol.*, **16**, 281–289.

Groothuis, T.G.G. and Meeuwissen, G. (1992). The influence of testosterone on the development and fixation of the form of displays in two age classes of young Black-headed gulls. *Anim. Behav.*, **43**, 189–208.

Groothuis, T.G.G., Müller, W. von Engelhardt, N., Carere, C., and Eising, C. (2005). Maternal androgens as a tool to adjust offspring development in avian species. *Neurosci. BioBehav. Rev.*, **29**, 329–352.

Hall, W.G. and Williams, C.L. (1983). Suckling isn't feeding, or is it? A search for developmental continuities. *Adv. Study Behav.*, **13**, 219–254.

Hamilton, W.D. (1964). The genetical evolution of social behavior (I and II). *J. Theor. Biol.*, **7**, 1–16 and 17–52.

Hammerstein, P. (1996). Darwinian adaptation, population genetics and the streetcar theory of evolution. *J. Math. Biol.*, **34**, 511–532.

Hampton, R.R. and Sherry, D.F. (1992). Food storing by Mexican chickadees and bridled titmice. *Auk*, **109**, 665–666.

Hampton, R.R. and Shettleworth, S.J. (1996a). Hippocampal lesions impair memory for location but not color in passerine birds. *Behav. Neurosci.*, **110**, 831–835.

Hampton, R.R. and Shettleworth, S.J. (1996b). Hippocampus and memory in a food-storing and in a nonstoring bird species. *Behav. Neurosci.*, **110**, 946–964.

Hampton, R.R., Sherry, D.F., Shettleworth, S.J., Khurgel, M., and Ivy, G. (1995). Hippocampal volume and food-storing behavior are related in parids. *Brain Behav. Evol.*, **45**, 54–61.

Hampton, R.R., Healy, S.D., Shettleworth, S.J., and Kamil, A.C. (2002). 'Neuroecologists' are not made of straw. *Trends Cogn. Sci.*, **6**, 6–7.

Harlow, H.F. (1958). The nature of love. *Amer. Psychologist*, **13**, 673–685.

Harlow, H.F. and Harlow, M.K. (1962). Social deprivation in monkeys. *Scient. Amer.*, **207**, 136–146.

Harvey, P.H. and Nee, S. (1997). The phylogenetic foundations of behavioural ecology. In *Behavioural Ecology. An Evolutionary Approach. 4th edn.* ed. J.R. Krebs and N.B. Davies. Oxford, UK: Blackwell Science Ltd., pp. 334–349.

Harvey, P.H. and Pagel M.D. (1991). *The Comparative Method in Evolutionary Biology.* Oxford, UK: Oxford University Press.

Hauser, M.D. (1996). *The Evolution of Communication.* Cambridge, MA: MIT Press.

Hauser, M.D., Chomsky, N., and Fitch, W.T. (2002). The faculty of language: what is it, who has it, and how did it evolve? *Science*, **298**, 1569–1579.

Healy, S. and Braithwaite, V. (2000). Cognitive ecology: a field of substance? *Trends Ecol. Evol.*, **15**, 22–26.

Healy, S.D. and Krebs, J.R. (1992). Delayed-matching-to-sample by marsh tits and great tits. *Q. J. Exp.Psychol. B – Comp.Physiol. Psychol.*, **45B**, 33–47.

Healy, S.D. and Krebs, J.R. (1992). Food storing and the hippocampus in corvids: amount and volume are correlated. *Proc. R. Soc. Lond. B*, **248**, 241–245.

Healy, S.D. and Krebs, K.R. (1993). Development of hippocampal specialization in a food-storing bird. *Behav. Brain Res.*, **53**, 127–131.

Healy, S.D., Clayton, N.S., and Krebs, J.R. (1994). Development of hippocampal specialization in two species of tit (*Parus* spp.). *Behav. Brain Res.*, **61**, 23–28.

Healy, S.D., de Kort, S.R., and Clayton, N.S. (2005). The hippocampus, spatial memory and food hoarding: a puzzle revisited. *Trends Ecol. Evol.*, **20**, 17–22.

Hebb, D.O. (1949). *The Organization of Behavior.* New York: John Wiley & Sons.

Hebb, D.O. (1953). Heredity and environment in animal behaviour. *Bri. J. Anim. Behav.*, **1**, 43–47.

Heidegger, M. (1996). *Being and Time*, ed. J. Stambaugh. Albany NY: Trans. State University of New York Press.

Heiligenberg, W. (1977). *Principles of Electrolocation and Jamming Avoidance in Electric Fish: A Neuroethological Approach*. Berlin: Springer.

Heinroth, O. (1909). Beobachtungen bei der zucht des ziegenmelkers (*Caprimulgus europaeus* L.). *J. Ornithol.*, **57**, 56–83.

Hennig, W. (1950). *Grundzuge einer Theorie der Phylogenetischen Systematik*. Berlin: Deutscher zentralverlag.

Hennig, W. (1966). *Phylogenetic Systematics*. Urbana: University of Illinois Press.

Hickok, G., Bellugi, U., and Klima, E.S. (1998). The neural organization of language: evidence from sign language aphasia. *Trends Cogn. Sci.*, **2**, 129–136.

Hilton, S.C. and Krebs, J.K. (1990). Spatial memory of four species of *Parus*: performance in an open-field analogue of a radial maze. *Q. J. Exp.Psychol.*, **4B**, 345–368.

Hinde, R.A. (1959). Unitary drives. *Anim. Behav.*, **7**, 130–141.

Hinde, R.A. (1960). Energy models of motivation. *Symp.Soc. Exp.Biol.*, **14**, 199–213.

Hinde, R.A. (1962). Some aspects of the imprinting problem. *Symp. Zool. Soc. Lond.*, **8**, 129–138

Hinde, R.A. (1970). *Animal Behavior, 2nd edn*. New York: McGraw-Hill.

Hinde, R.A. (1975). The concept of function. In *Function and Evolution in Behaviour*, ed. G.P. Baerends, C. Beer, and A. Manning. Oxford: Oxford University Press, pp. 2–15.

Hinde, R.A. (1977). Mother-infant separation and the nature of inter-individual relationships: experiments with rhesus monkeys. *Proc. Roy. Soc. Lond. B*, **196**, 29–50.

Hinde, R.A. (1982). *Ethology*. London: Fontana Paperbacks.

Hinde, R.A. and Spencer-Booth, Y. (1971). Effects of brief separation from mother on rhesus monkeys. *Science*, **173**, 111–118.

Hines, M., Golombok, S., Rust, J., Johnston, K.J., Golding, J., and the Avon Longitudinal Study of Parents and Children Study Team (2002). Testosterone during pregnancy and gender role behaviour of preschool children: a longitudinal population study. *Child Devel.*, **73**, 1678–1687.

Hogan, J.A. (1971). The development of a hunger system in young chicks. *Behaviour*, **39**, 128–201.

Hogan, J.A. (1988). Cause and function in the development of behavior systems. In *Handbook of Behavioral Neurobiology, Vol. 9*, ed. E.M. Blass. New York: Plenum Press, pp. 63–106.

Hogan, J.A. (1990). Animal behavior. In *Foundations of Psychology*, ed. J.E. Grusec, R.S. Lockhart, and G.C. Walters. Toronto: Copp Clark Pitman, pp. 138–186.

Hogan, J.A. (1994). The concept of cause in the study of behavior. In *Causal Mechanisms of Behavioural Development*, ed. J.A. Hogan and J.J. Bolhuis. Cambridge, UK: Cambridge University Press, pp. 3–15.

Hogan, J.A. (1997). Energy models of motivation: a reconsideration. *Appl. Anim. Behav. Sci.*, **53**, 89–105.

Hogan, J.A. (2001). Development of behavior systems. In *Handbook of Behavioral Neurobiology, Vol. 13: Developmental Psychobiology*, ed. E.M. Blass. New York: Kluwer Academic/Plenum, pp. 229–279.

Hogan, J.A. (2005). Causation: the study of behavioural mechanisms. *Anim. Biol.*, **55**, 323–341. (Chapter 3 in this volume).

Hogan, J.A. and Bolhuis, J.J. (2005). The development of behavior: trends since Tinbergen (1963). *Anim. Biol.*, **55**, 371–398. (Chapter 5 in this volume).

Hogan, J.A. and Ghirlanda, S. (2003). A quantitative test of dustbathing motivation. (Abstract). *Revista de Etologia*, **5**, 171.

Hogan, J.A. and Roper, T.J. (1978). A comparison of the properties of different reinforcers. *Adv. Study Behav.*, **8**, 155–255.

Hogan, J.A. and van Boxel, F. (1993). Causal factors controlling dustbathing in Burmese red junglefowl: some results and a model. *Anim. Behav.*, **46**, 627–635.

Hogan, J.A., Honrado, G.I., and Vestergaard, K.S. (1991). Development of a behavior system: dustbathing in the Burmese red junglefowl (*Gallus gallus spadiceus*): II. Internal factors. *J. Comp. Psychol.*, **105**, 269–273.

Hogan-Warburg, A.J. and Hogan, J.A. (1981). Feeding strategies in the development of food recognition in young chicks. *Anim. Behav.*, **29**, 143–154.

Hollis, K.L., ten Cate, C., and Bateson, P. (1991). Stimulus representation: A subprocess of imprinting and conditioning. *J. Comp. Psychol.*, **105**, 307–317.

Holmes, W.G. (2004). The early history of Hamiltonian-based research on kin recognition. *Annls Zool. Fenn.*, **41**, 691–711.

Holmes, W.G. and Sherman, P.W. (1983). Kin recognition in animals. *Am. Scient.*, **71**, 46–55.

Honey, R.C. and Bolhuis, J.J. (1997). Imprinting, conditioning, and within-event learning. *Q. J. Exp. Psychol.*, **50B**, 97–110.

Hopkins, B. and Butterworth, G. (1990). Concepts of causality in explanations of development. In *Causes of Development*, ed. G. Butterworth and P. Bryant. London: Harvester Wheatsheaf, pp. 3–32.

Horn, G. (1985). *Memory, Imprinting, and the Brain*. Oxford, UK: Clarendon Press.

Horn, G. (1998). Visual imprinting and the neural mechanisms of recognition memory. *Trends Neurosci.*, **21**, 300–305.

Horn, G. (2000). In memory. In *Brain, Perception, Memory. Advances in Cognitive Neuroscience*, ed. J.J. Bolhuis. Oxford, UK: Oxford University Press, pp. 329–363.

Horn, G. (2004). Pathways of the past: The imprint of memory. *Nature Rev. Neurosci.*, **5**, 108–120.

Hoshooley, J.S. and Sherry, D.F. (2004). Neuron production, neuron number, and structure size are seasonally stable in the hippocampus of the food-storing black-capped chickadee (*Poecile atricapillus*). *Behav. Neurosci.*, **118**, 345–355.

Hoshooley, J.S., Phillmore, L.S., and MacDougall-Shackleton, S.A. (2005). An examination of avian hippocampal neurogenesis in relationship to photoperiod. *NeuroReport*, **16**, 987–991.

Hoshooley, J.S., Phillmore, L.S., Sherry, D.F., and MacDougall-Shackleton, S.A. (2007). Annual cycle of the black-capped chickadee: seasonality of food-storing and the hippocampus. *Brain Behav. Evol.*, **69**, 161–168.

Houston, A.I. and McNamara, J.M. (1999). *Models of Adaptive Behaviour: an Approach Based on State*. Cambridge, UK: Cambridge University Press.

Houx, B.B. and ten Cate, C. (1998). Do contingencies with tutor behaviour influence song learning in zebra finches? *Behaviour*, **135**, 599–614.

Houx, B.B. and ten Cate, C. (1999). Song learning from playback in zebra finches: is there an effect of operant contingency? *Anim. Behav.*, **57**, 837–845.

Howdeshell, K.L. and vom Saal, F.S. (2000). Developmental exposure to bisphenol A: interaction with endogenous estradiol during pregnancy in mice. *Amer. Zool.*, **40**, 429–437.

Howdeshell, K.L., Hotchkiss, A.K., Thayer, K.A., Vandenbergh, J.G., Welshons, W.V., and vom Saal, F.S. (1999). Exposure to bisphenol A advances puberty. *Nature*, **401**, 763–764.

Hoyle, G. (1984). The scope of neuroethology. *Behav. Brain Sci.*, **7**, 367–384

Hughes, B.O. and Duncan, I.J.H. (1988). The notion of ethological 'need', models of motivation and animal welfare. *Anim. Behav.*, **36**, 1696–1707.

Hull, D.L. (1988). *Science as Progress*. Chicago: University of Chicago Press.

Hultsch, H. (1993). Tracing the memory mechanisms in the song acquisition of nightingales. *Neth. J. Zool.*, **43**, 155–171.

Hunt, S., Cuthill, I.C., Bennett, A.T.D., and Griffiths, R.M. (1999). Preference for ultraviolet partners in the blue tit. *Anim. Behav.*, **58**, 809–815.

Huxley, J.S. (1914). The courtship habits of the Great Crested Grebe (*Podiceps cristatus*); with an addition to the theory of sexual selection. *Proc. Roy. Soc. London*, (1914), 491–562.

Huxley, J.S. (1923). Courtship activities in the red-throated diver (*Colymbus stellatus* Pontopp); together with a discussion on the evolution of courtship in birds. *J. Linn. Soc.* **35**, 253–291.

Huxley, J.S. (1942). *Evolution: the Modern Synthesis*. London: Allen & Unwin.

Immelmann, K. (1972). The influence of early experience upon the development of social behaviour in estrildine finches. In *Proc. XV Intl. Ornithological Congress, Den Haag, 1970*, ed. K.H. Voous. Leiden: Brill, pp. 316–338.

Immelmann, K. (1979). Genetical constraints on early learning: a perspective from sexual imprinting in birds and other species. In *Theoretical. Advances in Behavior Genetics*, ed. J.R. Royce and P. Mos. Alphen aan de Rijn, NL: Sijthof & Noordhoff, pp. 121–137.

Immelmann, R., Lassek, R., Pröve, R., and Bischof, H.-J. (1991). Influence of adult courtship experience on the development of sexual preferences in zebra finch males. *Anim. Behav.*, **42**, 83–89.

Jackson, R.R. and Wilcox, R.S. (1998). Spider-eating spiders. *Amer. Scientist*, **86**, 350–357.

Jackson, R.R., Pollard, S.D., and Cerveira, A.M. (2002). Opportunistic use of cognitive smokescreens by araneophagic jumping spiders. *Anim. Cogn.*, **5**, 147–157.

Jalles-Filho, E., Grassetto, R., da Cunha, T., and Salm, R.A. (2001). Transport of tools and mental representation: is capuchin monkey tool behaviour a useful model of Plio-Pleistocene hominid technology? *J. Hum. Evol.*, **40**, 365–377.

James, W. (1890). *The Principles of Psychology*. New York: Holt.

Jamieson, I.G. and Craig, J.L. (1987). Critique of helping behavior in birds: a departure from functional explanations. In *Perspectives in Ethology, vol. 7*, ed. P.P.G. Bateson and P.H. Klopfer. New York: Plenum Press, pp. 79–98.

Jarvis, E.D. and Nottebohm, F. (1997). Motor-driven gene expression. *Proc. Natl Acad. Sci. USA*, **94**, 4097–4102.

Jennions, M.D. and Møller, A.P. (2002). Relationships fade with time: a meta-analysis of temporal trends in publication in ecology and evolution *Proc. R. Soc. Lond. B*, **269**, 43–48.

Jennions, M.D. and Møller, A.P. (2003). A survey of the statistical power of research in behavioral ecology and animal behavior. *Behav. Ecol.*, **14**, 438–445.

Johnsen, P.F., Vestergaard, K.S., and Nørgaard-Nielsen, G. (1998). Influence of early fearing conditions on the development of feather pecking and cannibalism in domestic fowl. *Appl. Anim. Behav. Sci.*, **60**, 25–41.

Johnson, M.H. and Bolhuis, J.J. (2000). Predispositions in perceptual and cognitive development. In *Brain, Perception, Memory. Advances in Cognitive Neuroscience*, ed. J.J. Bolhuis. Oxford, UK: Oxford University Press, pp. 69–84.

Johnson, M.H. and Horn, G. (1988). Development of filial preferences in dark-reared chicks. *Anim. Behav.*, **36**, 675–683.

Johnson, M.H., Davies, D.C., and Horn, G. (1989). A sensitive period for the development of a predisposition in dark-reared chicks. *Anim. Behav.*, **37**, 1044–1046.

Jones, D. Gonzalez-Lima, F., Crews, D., Galef, B.G., and Clark, M.M. (1997). Effects of intrauterine position on the metabolic capacity of the hypothalamus of female gerbils: a cytochrome oxidase study. *Physiol. Behav.*, **61**, 513–519.

Jusczyk, P.W. (1997). Finding and remembering words: some beginnings by English-learning infants. *Curr. Directions Psychol. Sci.*, **6**, 170–174.

Kacelnik, A. and Cuthill, I.C. (1987). Starlings and optimal foraging theory: modelling in a fractal world. In *Foraging Behavior*, ed. A. Kamil, J.R. Krebs, and H.R. Pulliam New York: Plenum, pp. 303–333.

Kaiser, S. and Sachser, N. (2005). The effects of prenatal social stress on behaviour: mechanisms and function. *Neurosci. Behav. Revs.*, **29**(2), 283–294.

Kako, E. (1999). Elements of syntax in the systems of three language-trained animals. *Anim. Learning Behav.* **27**, 1–14.

Kamil, A.C. (1988). A synthetic approach to the study of animal intelligence. In *Comparative Perspectives on Modern Psychology*, ed. D.W. Leger. Nebraska: University of Nebraska Press, pp. 230–257.

Kamil, A.C. and Balda, R.P. (1985). Cache recovery and spatial memory in Clark's nutcracker (*Nucifraga columbiana*). *J. Exp.Psychol.: Anim. Behav. Proc..*, **11**, 95–111.

Kamil, A.C., Balda, R.P., and Olson, D.J. (1994). Performance of 4 seed-caching corvid species in the radial-arm maze analog. *J. Comp.Psychol.*, **108**, 385–393.

Karmiloff-Smith, A. (1992). *Beyond Modularity: A Developmental Perspective on Cognitive Science*. Cambridge, MA: MIT Press.

Karmiloff-Smith, A. (1998). Development itself is the key to understanding developmental disorders. *Trends Cogn. Sci.*, **2**, 389–398.

Killeen, P.R. (2001). The four causes of behavior. *Curr. Dir. Psychol. Sci.*, **10**, 136–140.

Kilner, R.M. and Davies, N.B. (1999). Signals of need in parent–offspring communication and their exploitation by the common cuckoo. *Nature*, **397**, 667–672.

Kirkpatrick, M. (1982). Sexual selection and the evolution of female choice. *Evolution*, **36**, 1–12.

Kirn, J., O'Loughlin, B., Kasparian, S., and Nottebohm, F. (1994). Cell death and neuronal recruitment in the high vocal center of adult male canaries are temporally related to changes in song. *Proc. Natl Acad. Sci. USA*, **91**, 7844–7848.

Kirn, J.R., Clower, R.P., Kroodsma, D.E., and DeVoogd, T.J. (1989). Song-related brain regions in the red winged blackbird are affected by sex and season but not repertoire size. *J. Neurobiol.*, **20**, 139–163.

Konishi, M. (1965). The role of auditory feedback in the control of vocalization in the white-crowned sparrow. *Z. Tierpsychol.*, **22**, 770–783.

Konishi, M. (1993). Listening with two ears. *Scient. Am.*, **268**, 66–73.

Konishi, M. (1995). Neural mechanisms of auditory image formation. In *The Cognitive Neurosciences*, ed. M.S. Gazzaniga. Cambridge MA: MIT Press, pp. 269–277.

Konishi, M. (2003). Coding of auditory space. *Annu. Rev. Neurosci.*, **26**, 31–55.

Kraemer, G.W. (1992). A psychobiological theory of attachment. *Behav. Brain Sci.*, **15**, 493–511.

Kramer, G. (1953). Wird die Sonnenhöhe bei der Heimfindeorientierung verwertet? *J. Ornithol.*, **94**, 201–219.

Krebs, J.R. and Davies, N.B. (1993). *An Introduction to Behavioural Ecology*. 3rd edn. Oxford, UK: Blackwell Science Ltd.

Krebs, J.R. and Davies, N.B. (1997). The evolution of behavioural ecology. In *Behavioural Ecology. An Evolutionary Approach*. 4th edn. Oxford, UK: Blackwell Scientific, pp. 3–12.

Krebs, J.R., Sherry, D.F., Healy, S.D., Perry, V.H., and Vaccarino, A.L. (1989). Hippocampal specialization of food-storing birds. *Proc. Natl Acad. Sci., USA*, **86**, 1388–1392.

Krebs, J.R., Erichsen, J.T., and Bingman, V.P. (1991). The distribution of neurotransmitters and neurotransmitter-related enzymes in the dorsomedial telencephalon of the pigeon (*Columba livia*). *J. Comp.Neurol.*, **314**, 467–477.

Krebs, J.R., Clayton, N.S., Hampton, R.R., and Shettleworth, S.J. (1995). Effects of photoperiod on food-storing and the hippocampus in birds. *NeuroReport*, **6**, 1701–1704.

Krebs, J.R., Clayton, N.S., Healy, S.D., Cristol, D.A., Patel, S.N., and Joliffe, A.R. (1996). The ecology of the brain: food-storing and the hippocampus. *Ibis*, **138**, 34–46.

Kruijt, J.P. (1964). Ontogeny of social behavior in Burmese red junglefowl (*Gallus gallus spadiceus*). *Behaviour Suppl.* **9**, 1–201.

Kruijt, J.P. (1985). On the development of social attachments in birds. *Neth. J. Zool.*, **35**, 45–62.

Kruijt, J.P. and Meeuwissen, G.B. (1991). Sexual preferences of male zebra finches: effects of early adult experience. *Anim. Behav.*, **42**, 91–102.

Kruijt, J.P. and Meeuwissen, G.B. (1993). Consolidation and modification of sexual preferences in adult male zebra finches. *Neth. J. Zool.*, **43**, 68–79.

Kruijt, J.P., Bossema, I., and Lammers, G.J. (1982). Effect of early experience and male activity on mate choice in mallard females (*Anas platyrhynchos*). *Behaviour*, **80**, 32–43.

Kruijt, J.P., ten Cate, C., and Meeuwissen, G.B. (1983). The influence of siblings on the development of sexual preferences of male zebra finches. *Devel. Psychobiol.*, **16**, 233–239.

Kruuk, H. (2003). *Niko's Nature. A Life of Niko TInbergen and His Science of Animal Behaviour*. Oxford, UK: Oxford University Press.

Kuhl, P.K. (1994). Learning and representation in speech and language. *Curr. Opin. Neurobiol.*, **4**, 812–822.

Kuhl, P.K., Williams, K.A., Lacerda, R., Stevens, K.N., and Lindblom, B. (1992). Linguistic experience alters phonetic perception in infants by 6 months of age. *Science*, **255**, 606–608.

Laland, K.N. and Brown, G.R. (2002). *Sense and Nonsense. Evolutionary Perspectives on Human Behaviour*. Oxford, UK: Oxford University Press.

Larsen, B.H., Vestergaard, K.S., and Hogan, J.A. (2000). Development of dustbathing behavior sequences in the domestic fowl: the significance of functional experience. *Devel. Psychobiol.*, **37**, 5–12.

Lashley, K.S. (1938). Experimental analysis of instinctive behavior. *Psychol. Rev.*, **45**, 445–471.

Lashley, K.S. (1951). The problem of serial order in behavior. In *Cerebral Mechanisms in Behavior*, ed. L.A. Jeffress New York: Wiley, pp. 112–136.

Lavenex, P., Steele, M.A., and Jacobs, L.F. (2000a). The seasonal pattern of cell proliferation and neuron number in the dentate gyrus of wild adult eastern gray squirrels. *Eur. J. Neurosci.*, **12**, 643–648.

Lavenex, P., Steele, M.A., and Jacobs, L.F. (2000b). Sex differences, but no seasonal variations in the hippocampus of foodcaching squirrels: a stereological study. *J. Comp.Neurol.*, **425**, 152–166.

Lefebvre, L. and Bolhuis, J.J. (2003). Positive and negative correlates of feeding innovations in birds: evidence for limited modularity. In *Animal Innovation*, ed. S.M. Reader and K.N. Laland. Oxford, UK: Oxford University Press, pp. 39–61.

Lefebvre, L., Nicolakakis, N., and Boire, D. (2002). Tools and brains in birds. *Behaviour*, **139**, 939–973.

Lehrman, D.S. (1953). A critique of Konrad Lorenz's theory of instinctive behavior. *Q. Rev. Biol.*, **28**, 337–363.

Lehrman, D.S. (1965). Interaction between internal and external environments in the regulation of the reproductive cycle of the ring dove. In *Sex and Behavior*, ed. F.A. Beach. New York: Wiley, pp. 355–380.

Lehrman, D.S. (1970). Semantic and conceptual issues in the nature–nurture problem. In *Development and Evolution of Behavior*, ed. L.R. Aronson, E. Tobach, D.S. Lehrman, and J.S. Roisenblatt. San Francisco: Freeman, pp. 17–52.

Leitner, S. and Catchpole, C.K. (2004). Syllable repertoire and the size of the song control system in captive canaries (*Serinus canaria*). *J. Neurobiol.*, **60**, 21–27.

Leitner, S., Voigt, C., Garcia-Segura, L.-M., Van't Hof, T., and Gahr, M. (2001). Seasonal activation and inactivation of song motor memories in wild canaries is not reflected in neuroanatomical changes of forebrain song areas. *Horm. Behav.*, **40**, 160–168.

Lenneberg, E.H. (1967). *Biological Foundations of Language*. New York: John Wiley.

Li, D., Jackson, R.R., and Lim, M.L.M. (2003). Influence of background and prey orientation on an ambushing predator's decisions. *Behaviour*, **140**, 739–764.

Licht, P., Frank, L.G., Pavgi, S., Yalcinkaya, T.M., Siteri, P.K., and Glickman, S.E. (1992). Hormonal correlates of "masculinization" in female spotted hyenas: 2, Maternal and fetal steroids. *J. Reprod. Fertil.*, **95**, 463–474.

Lipar, J.L. and Ketterson, E.D. (2000). Maternally derived yolk testosterone enhances the development of the hatching muscle in the red-winged blackbird. *Proc. Roy. Soc. Lond. B*, **267**, 2005–2010.

Locke, J.L. (1993). *The Child's Path to Spoken Language*. Cambridge, MA: Harvard University Press.

Locke, J.L. (1994). The biological building blocks of spoken language. In *Causal Mechanisms of Behavioural Development*, ed. J.A. Hogan and J.J. Bolhuis. Cambridge, UK: Cambridge University Press, pp. 300–324.

Locke, J.L. and Snow, C. (1997). Social influences on vocal learning in human and nonhuman primates. In *Social Influences on Vocal Development*, ed. C.T. Snowdon and M. Hausberger. Cambridge, UK: Cambridge University Press, pp. 274–292.

Logan, C.A. (1983). Biological diversity in avian vocal learning. In *Advances in Analysis of Behavior, Vol. 3: Biological Factors in Learning*, ed. M.D. Zeiler and P. Harzem Chichester, UK: Wiley, pp. 143–176.

Lohmann, K.J. and Johnsen, S. (2000). The neurobiology of magnetoreception in vertebrate animals. *Trends Neurosci.*, **23**, 153–159.

Lorenz, K. (1935). Der Kumpan in der Umwelt des Vogels. *J. Ornithol.*, **83**, 137–213, 289–413. [Translation: Companions as factors in the bird's environment. In *Studies in Animal and Human Behavior, vol. 1*, 1970. London: Methuen, pp. 101–258.]

Lorenz, K. (1937). Über die Bildung des Instinktbegriffes. *Naturwissenschaften*, **25**, 289–300, 307–318, 324–331. [Translation: The establishment of the instinct concept. In *Studies in Animal and Human Behavior, vol. 1*, 1970. London: Methuen, pp. 259–315.]

Lorenz, K. (1941). Comparative studies of the motor patterns of the Anatinae (translated by R. Martin 1971). *Studies in Animal and Human Behavior*, **2**, 14–18 and 106–114.

Lorenz, K.Z. (1950). The comparative method in studying innate behaviour patterns. *Symp. Soc. Exp. Biol.*, **4**, 221–268.

Lorenz, K.Z. (1965). *Evolution and Modification of Behaviour*. Chicago: University of Chicago Press.

Lucas, J.R., Brodin, A., de Kort, S.R., and Clayton, N.S. (2004). Does hippocampal size correlate with the degree of caching specialization? *Proc. Roy. Soc. Lond. B*, **271**, 2423–2429.

MacArthur, R.H. and Pianka, E.R. (1966). On optimal use of a patchy environment. *Am. Nat.*, **100**, 603–609.

MacDougall-Shackleton, S.A. and Ball, G.F. (1999). Comparative studies of sex differences in the song-control system of songbirds. *Trends Neurosci.*, **22**, 432–436.

MacDougall-Shackleton, S.A. and Ball, G.F. (2002). Revising hypotheses does not indicate a flawed approach. *Trends Cogn. Sci.*, **6**, 68–69.

MacDougall-Shackleton, S.A., Hulse, S.H., and Ball, G.F. (1998a). Neural bases of song preferences in female zebra finches (*Taeniopygia guttata*). *NeuroReport*, **9**, 3047–3052.

MacDougall-Shackleton, S.A., Hulse, S.H., and Ball, G.F. (1998b). Neural correlates of singing behavior in male zebra finches (*Taeniopygia guttata*). *J. Neurobiol.*, **36**, 421–430.

MacDougall-Shackleton, S.A., Sherry, D.F., Clark, A.P., Pinkus, R., and Hernandez, A.M. (2003). Photoperiodic regulation of food-storing and hippocampus volume in black-capped chickadees (*Poecile atricapilla*). *Anim. Behav.*, **65**, 805–812.

Macphail, E.M. and Bolhuis, J.J. (2001). The evolution of intelligence: adaptive specialisations versus general process. *Biol. Rev.*, **76**, 341–364.

Maney, D.L., MacDougall-Shackleton, E.A., MacDougall-Shackleton, S.A., Ball, G.F., and Hahn, T.P. (2003). Immediate early gene response to hearing song correlates with receptive behavior and depends on dialect in female white-crowned sparrows. *J. Comp. Physiol. A*, **189**, 667–674.

Manning, A. and Dawkins, M.S. (1995). *An Introduction to Animal Behaviour*. Cambridge, UK: Cambridge University Press.

Manning, C.J., Wakeland, E.K., and Potts, W.K. (1992). Communal nesting patterns in mice implicate MHC genes in kin recognition. *Nature*, **360**, 581–583.

Margoliash, D. (1986). Preference for autogenous song by auditory neurons in a song system nucleus of the white-crowned sparrow. *J. Neurosci.*, **6**, 1643–1661.

Margoliash, D. and Konishi, M. (1985). Auditory representation of autogenous song in the song system of white-crowned sparrows. *Proc. Natl Acad. Sci. USA*, **82**, 5997–6000.

Marler, P. (1970a). A comparative approach to vocal learning: Song development in white-crowned sparrows. *J. Comp. Physiol. Psychol.* (monograph supplement), **71**, 1–25.

Marler, P. (1970b). Birdsong and speech development: could there be parallels? *Amer. Scientist*, **58**, 669–673.

Marler, P. (1976). Sensory templates in species-specific behavior. In *Simpler Networks and Behavior*, ed. J.C. Fentress. Sunderland, MA: Sinauer, pp. 314–329.

Marler, P. (1984). Song learning: innate species differences in the learning process. In *The Biology of Learning*. ed. P. Marler and H.S. Terrace. Dahlem Workshop Reports-Life Sciences Research Report *29*. Berlin: Springer.

Marler, P. (1987). Sensitive periods and the roles of specific and general sensory stimulation in birdsong learning. In *Imprinting and Cortical Plasticity. Comparative Aspects of Sensitive Periods*, ed. J.P. Rauschecker and P. Marler. New York: John Wiley and Sons, pp. 99–135.

Marler, P. (1991). Song-learning behavior: the interface with neuroethology. *Trends Neurosci.*, **14**, 199–206.

Marler, P. (1997). Three models of song learning: evidence from behavior. *J. Neurobiol.*, **33**, 501–516.

Marler, P. and Doupe, A. (2000). Singing in the brain. *Proc. Natl Acad. Sci. USA*, **97**, 2965–2967.

Marler, P. and Peters, S. (1982). Subsong and plastic song: their role in the vocal learning process. In *Acoustic communication in birds*, Vol. 2, ed. D.E. Kroodsma and E.H. Miller. New York: Academic Press, pp. 25–50.

Marler, P. and Peters, S.S. (1989). Species differences in auditory responsiveness in early vocal learning. In *The Comparative Psychology of Audition: Perceiving Complex Sounds*, ed. S. Hulse and R. Dooling. Hillsdale, NJ: Lawrence Erlbaum, pp. 243–273.

Marr, D. (1982). *Vision: A Computational Investigation into the Human Representation and Processing of Visual Information*. San Francisco, CA: Freeman.

Martins, E. (1996). *Phylogenies and the Comparative Method in Animal Behavior*. Oxford, UK: Oxford University Press.

Maynard Smith, J. (1982). *Evolution and the Theory of Games*. Cambridge, UK: Cambridge University Press.

Maynard Smith, J. and Price, G.R. (1973). The logic of animal conflict. *Nature*, **246**, 15–18.

Maynard Smith, J. and Reichart, S.E. (1984). A conflicting tendency model of spider agonistic behaviour in hybrid-pure line comparison. *Anim. Behav.*, **32**, 564–578.

Mayr, E. (1961). Cause and effect in biology. *Science*, **134**, 1501–1506.

Mayr, E. (1988). The multiple meanings of teleological. In *Toward a New Philosophy of Biology*. Cambridge, MA: Harvard University Press, pp. 38–66.

Mayr, E. (1993). Proximate and ultimate causations. *Biol. Philos.*, **8**, 93–94.

McClelland, J.L. and Rumelhart, D.E., Eds. (1985). *Parallel Distributed Processing: Explorations in the Microstructure of Cognition*. Vol. 2: *Psychological and Biological Models*. Cambridge, MA: MIT Press.

McFarland, D.J., (ed.) (1974). *Motivational Control Systems Analysis*. London: Academic Press.

McFarland, D.J. (1989). *Problems of Animal Behaviour*. London: Longman Scientific and Technical.

Meaney, M.J. ( 2001). Maternal care, gene expression and the transmission of individual differences in stress reactivity across generations. *Ann. Rev. Neurosci.*, **24**, 161–192.

Medawar, P.B. (1967). *The Art of the Soluble*. London: Methuen.

Mello, C.V. (2002). Mapping vocal communication pathways in birds with inducible gene expression. *J. Comp.Physiol. A*, **188**, 943–959,

Mello, C.V. and Clayton, D.F. (1994). Song-induced ZENK gene expression in auditory pathways of songbird brain and its relation to the song control system. *J. Neurosci.*, **14**, 6652–6666.

Mello, C.V., Vicario, D.S., and Clayton, D.F. (1992). Song presentation induces gene-expression in the songbird forebrain. *Proc. Natl Acad. Sci. USA*, **89**, 6818–6822.

Mercader, J., Panger, M., and Boesch, C. (2002). Excavation of a chimpanzee stone tool site in the African rainforest. *Science*, **296**, 1452–1455.

Milinski, M., Griffiths, S., Wegner, K.M., Reusch, T.B.H., Haas-Assenbaum, A., and Boehm, T. (2005). Mate choice decisions of stickleback females predictably modified by MHC peptide ligands. *Proc. Natl Acad. Sci. USA*, **102**, 4414–4418.

Moiseff, A. and Konishi, M. (1981). Neuronal and behavioral sensitivity to binaural time differences in the owl. *J. Neurosci.*, **1**, 40–48.

Moiseff, A. and Konishi, M. (1983). Binaural characteristics of units in the owl's brainstem auditory pathway: precursors of restricted spatial receptive fields. *J. Neurosci.*, **3**, 2553–2562.

Mook, D.G. (1983). In defense of external invalidity. *Am. Psych.*, **38**, 379–389.

Moore, C.L. (1995). Maternal contributions to mammalian reproductive development and divergence of males and females. In *Advances in the Study of Behavior*, ed. P.J.P. Slater, J.S. Rosenblatt, C.T. Snowdon, and M. Milinski. New York: Academic Press, pp. 47–118.

Morris, D. (1979). *Animal Days*. London: Jonathan Cape.

Mueller, W., Eising, C.M., Dijkstra, C., and Groothuis, T.G.G. (2004). Within-clutch patterns of yolk testosterone varies with the onset of incubation in black-headed gulls. *Behav. Ecol. Sociobiol.*, **15**, 893–897.

Munn, N.L. (1950). *Handbook of Psychological Research on the Rat*. Boston, MA: Houghton Mifflin.

Murray, E.A. (1996). What have ablation studies told us about the neural substrates of stimulus memory? *Sem. Neurosci.*, **8**, 13–22.

Nealen, P.M. (2005). An interspecific comparison using immuno-fluorescence reveals that synapse density in the avian song system is related to sex but not to male song repertoire size. *Brain Res.*, **1032**, 50–62.

Neff, B.D. and Pitcher, T.E. (2005). Genetic quality and sexual selection: an integrated framework for good genes and compatible genes. *Mol. Ecol.*, **14**, 19–38.

Nelson, D. (1997). Social interaction and sensitive phases for song learning: a critical review. In *Social Influences on Vocal Development*, ed. C.T. Snowdon and M. Hausberger. Cambridge, UK: Cambridge Univ. Press, pp. 7–22.

Nelson, D.A. and Marler, P. (1990). The perception of birdsong and an ecological concept of signal space. In *Comparative Perception: Complex Signals*, ed. W.C. Stebbins and M.A. Berkeley. New York: John Wiley & Sons, pp. 443–478.

Nelson, K. (1965). After-effects of courtship in the male three-spined stickleback. *Z. Vgl. Physiol.*, **50**, 569–597.

Nelson, R.J. (1999). *An Introduction to Behavioral Endocrinology*, 2nd edn. Sunderland, MA: Sinauer.

Newman, S.W. (1999). The medial extended amygdala in male reproductive behavior: a node in the mammalian social behavior network. *Ann. NY Acad. Sci.*, **877**, 242–257.

Nicol, C.J., Lindberg, A.C., Phillips, A.J., Pope, S.J, Wilkins, L.J., and Green, L.E. (2001). Influence of prior exposure to wood shavings on feather pecking, dustbathing and foraging in adult laying hens. *Appl. Anim. Behav Sci.*, **73**, 141–155.

Nordeen, K.W. and Nordeen, E.J. (1997). Anatomical and synaptic substrates for avian song learning. *J. Neurobiol.*, **33**, 532–548.

Nordeen, K.W., Marler, P., and Nordeen, E.J. (1989). Addition of song-related neurons in swamp sparrows coincides with memorization, not production, of learned songs. *J. Neurobiol.* **20**, 651–661.

Nottebohm, F. (1981). A brain for all seasons: cyclical anatomical changes in song control nuclei of the canary brain. *Science*, **214**, 1368–1370.

Nottebohm, F. (1984). Birdsong as a model in which to study brain processes related to learning. *Condor*, **87**, 227–236.

Nottebohm, F. (2000). The anatomy and timing of vocal learning in birds. In *The Design of Animal Communication*, ed. M.D. Hauser and M. Konishi. Cambridge, MA: MIT Press, pp. 63–110.

Nottebohm, F. (2002). Neuronal replacement in adult brain. *Brain Res. Bull.*, **57**, 737–749.

Nottebohm, F. (2002). Why are some neurons replaced in adult brain? *J. Neurosci.*, **22**, 624–628.

Nottebohm, F., Stokes, T., and Leonard, C.M. (1976). Central control of song in the canary. *J. Comp.Neurol.*, **165**, 457–486.

Nottebohm, F., Kasparian, S., and Pandazis, C. (1981). Brain space for learned task. *Brain Res.*, **213**, 99–109.

Nottebohm, F., Alvarez-Buylla, A., Cynx, J. *et al.* (1990). Song learning in birds: the relation between perception and production. *Phil. Trans. Roy. Soc. Lond, B*, **329**, 115–124.

Nottebohm, R. (1991). Reassessing the mechanisms and origins of vocal learning in birds. *Trends Neurosci.*, **14**, 206–211.

Oppenheim, R.W. (1981). Ontogenetic adaptations and retrogressive processes in the development of the nervous system and behaviour: a neuroembryological perspective. In *Maturation and Development: Biological and Psychological Perspectives*, ed. K.J. Connolly and H.F.R. Prechtl. Philadelphia: Lippincott, pp. 73–109.

Oppenheim, R.W. (2001). Early development of behavior and the nervous system, an embryological perspective. In *Handbook of Behavioral Neurobiology, Vol. 13: Developmental Psychobiology*, ed. E.M. Blass. New York: Kluwer Academic/ Plenum, pp. 15–52.

Orzack, S.H. and Sober, E. (2001). Adaptation, phylogenetic inertia, and the method of controlled comparisons. In *Adaptationism and Optimality*, ed. S.H. Orzack and E. Sober. Cambridge, UK: Cambridge University Press, pp. 45–63.

Oyama, S. (1985). *The Ontogeny of Information*. Cambridge, UK: Cambridge University Press.

Palmer, A.R. (2000). Quasireplication and the contract of error: lessons from sex ratios, heritabilities and fluctuating asymmetry. *Annu. Rev. Ecol. Syst.*, **31**, 441–480.

Peña, J.L., Viete, S., Funabiki, K., Saberi, K., and Konishi, M. (2001). Cochlear and neural delays for coincidence detection in owls. *J. Neurosci.*, **21**, 9455–9459.

Penn, D. and Potts, W. (1998). MHC-disassortiative mating preferences reversed by cross-fostering. *Proc. Roy. Soc. Lond. B*, **265**, 1299–1306.

Penn, D.J. (2002). The scent of genetic compatibility: sexual selection and the major histocompatibility complex. *Ethology*, **108**, 1–21.

Penn, D.J. and Potts, W.K. (1999). The evolution of mating preferences and the major histocompatibility complex. *Am. Nat.*, **153**, 146–164.

Petherick, J.C., Seawright, E., Waddington, D., Duncan, I.J.H., and Murphy, L.B. (1995). The role of perception in the causation of dustbathing behavior in domestic fowl. *Anim. Behav.*, **49**, 1521–1530.

Petitto, L.A. and Marentette, P.F. (1991). Babbling in the manual mode: Evidence for the ontogeny of language. *Science*, **251**, 1493–1496.

Petrinovich, L. (1985). Factors influencing song development in the white-crowned sparrow (*Zonotrichia leucophrys*). *J. Comp.Psychol.*, **99**, 15–29.

Phelps, S.M. and Ryan, M.J. (1998). Neural networks predict response biases in female túngara frogs. *Proc. Roy. Soc., Lond. Ser. B*, **265**, 279–285.

Phelps, S.M. and Ryan, M.J. (2000). History influences signal recognition: neural network models of túngara frogs. *Proc. Roy. Soc., Lond. Ser. B*, **267**, 1633–1639.

Phelps, S.M., Ryan, M.J., and Rand, A.S. (2001). Vestigial preference functions in neural networks and tungara frogs. *Proc. Natl Acad. Sci. USA*, **98**, 13161–13166.

Pierce, G.J. and Ollason, J.G. (1987). Eight reasons why optimal foraging theory is a complete waste of time. *Oikos*, **49**, 111–118.

Pinker, S. (1994). *The Language Instinct*. London: Penguin Books.

Pinker, S. (1997). *How the Mind Works*. New York: W.W. Norton.

Pittendrigh, C.S. (1958). Adaptation, natural selection, and behavior. In *Behavior and Evolution*, ed. A. Roe and G.G. Simpson. New Haven: Yale University Press, pp. 390–416.

Podos, J. (1996). Motor constraints on vocal developement in a songbird. *Anim. Behav.*, **51**, 1061–1070.

Povinelli, D.J., Bering, J.M., and Giambrone, S. (2000). Toward a science of other minds: Escaping the argument by analogy. *Cogn. Sci.*, **24**, 509–541.

Pravosudov, V.V. (2003). Long term moderate elevation of corticosterone facilitates avian food-caching behavior and enhances spatial memory. *Proc. Roy. Soc. Lond. B*, **270**, 2599–2604.

Pravosudov, V.V. and Clayton, N.S. (2001). Effects of demanding foraging conditions on cache retrieval accuracy in foodcaching mountain chickadees (*Poecile gambeli*). *Proc. Roy. Soc. Lond. B*, **268**, 363–368.

Pravosudov, V.V. and Clayton, N.S. (2002). A test of the adaptive specialization hypothesis: population differences in caching, memory, and the hippocampus in black-capped chickadees (*Poecile atricapilla*). *Behav. Neurosci.* **116**, 513–522.

Pravosudov, V.V. and Omanska, A. (2005). Prolonged moderate elevation of corticosterone does not affect hippocampal anatomy or cell proliferation rates in mountain chickadees (*Poecile gambeli*). *J. Neurobiol.*, **62**, 82–91.

Pravosudov, V.V., Kitaysky, A.S., Wingfield, J.C., and Clayton, N.S. (2001). Long-term unpredictable foraging conditions and physiological stress response in mountain chickadees (*Poecile gambeli*). *Gen. Comp. Endocrinol.*, **123**, 324–331.

Pravosudov, V.V., Lavenex, P., and Clayton, N.S. (2002). Changes in spatial memory mediated by experimental variation in food supply do not affect hippocampal anatomy in mountain chickadees (*Poecile gambeli*). *J. Neurobiol.* **51**, 142–148.

Pulliam, H.R. (1974). On the theory of optimal diets. *Am. Nat.*, **108**, 59–74.

Putz, O. and Crews, D. (2006). Embryonic origin of mate choice in a lizard with temperature-dependent sex determination. *Devel. Psychobiol.*, **48**, 29–38.

Raby, C.R., Alexis, D.M., Dickinson, A., and Clayton, N.S. (2007). Planning for the future by western scrub-jays. *Nature*, **445**, 919–921.

Rand, A.S. and Ryan M.J. (1981). The adaptive significance of a complex vocal repertoire in a neotropical frog (*Physalaemus pustulosus*). *Z. Tierpsychol.*, **57**, 209–214.

Rand, A.S., Ryan, M.J., and Wilczynski, W. (1992). Signal redundancy and receiver permissiveness in acoustic mate recognition by the túngara frog *Physalaemus pustulosus*. *Am. Zool.*, **32**, 81–90.

Rauschecker, J.P. and Marler, P. (eds.) (1987). *Imprinting and Cortical Plasticity*. New York: Wiley.

Real, L.A. (1993). Toward a cognitive ecology. *Trends Ecol. Evol.*, **8**, 413–417.

Reeve, H.K. and Sherman, P.A. (1993). Adaptation and the goals of evolutionary research. *Q. Rev. Biol.*, **68**, 1–32.

Reeve, H.K. and Sherman, P.A. (2001). Optimality and phylogeny: a critique of current thought. In *Adaptationism and Optimality*, ed. S. Orzack and E. Sober. Cambridge, UK: Cambridge University Press.

Reznick, D.A., Bryga, H., and Endler, J.A. (1990). Experimentally induced life-history evolution in a natural population. *Nature*, **346**, 357–359.

Rhen, T. and Crews, D. (2002). Variation in reproductive behavior within a sex; neural systems and endocrine activation. *J. Neuroendocr.*, **14**, 517–532.

Ribeiro, S., Cecchi, G.A., Magnasco, M.O., and Mello, C.V. (1998). Toward a song code: evidence for a syllabic representation in the canary brain. *Neuron*, **21**, 359–371.

Ridley, M. (1983). *The Explanation of Organic Diversity.the Comparative Method and Adaptations for Mating*. Oxford, UK: Clarendon Press.

Riebel, K. (2000). Early exposure leads to repeatable preferences for male song in female zebra finches. *Proc. Roy. Soc. Lond, B.*, **267**, 2553–2558.

Riebel, K. (2003a). Developmental influences on auditory perception in female zebra finches – is there a sensitive phase for song preference learning? *Anim. Biol.*, **53**, 73–87.

Riebel, K. (2003b). The 'mute' sex revisited: vocal production and perception learning in female song birds. *Adv. Study Behav.*, **33**, 49–86.

Riebel, K., Smallegange, I.M., Terpstra, N.J., and Bolhuis, J.J. (2002). Sexual equality in zebra finch song preference: evidence for a dissociation between song recognition and production learning. *Proc. Roy. Soc. Lond., B*, **269**, 729–733.

Riedstra, B. and Groothuis, T.G.G. (2002). Early feather pecking as a form of social exploration: the effect of group stability on feather pecking and tonic immobility in domestic chicks. *Appl. Anim. Behav. Sci.*, **77**, 127–138.

Ristau, C.A. (1991). *Cognitive Ethology: The Minds of Other Animals*. Hillsdale, New Jersey: Lawrence Erlbaum Associates.

Robinson, G.E. and Ben-Shahar, Y. (2002). Social behaviour and comparative genomics: new genes or new regulation? *Genes Brain Behav.*, **1**, 197–203.

Robinson, M.H. (1991). Niko Tinbergen, comparative studies and evolution. In *The Tinbergen Legacy*, ed. M.S. Dawkins, T.R. Halliday, and R. Dawkins. London: Chapman & Hall, pp. 100–128.

Röell, D.R. (2000). *The World of Instinct: Niko Tinbergen and the Rise of Ethology in the Netherlands (1920–50)*. Assen, Netherlands: Van Gorcum.

Rogers, L.J. and Deng, C. (2002). Factors affecting the development of lateralization in chicks. In *Comparative Vertebrate Lateralization*, ed. L.J. Rogers and R.J. Andrew. New York: Cambridge University Press, pp. 206–246.

Ros, A.F.H. (1999). Effects of testosterone on growth, plumage pigmentation and mortality in black-headed gull chicks. *Ibis*, **141**, 451–459.

Ros, A.F.H., Dieleman, St. J., and Groothuis, T.G.G. (2002). Social stimuli, testosterone, and aggression in gull chicks: support for the challenge. *Horm. Behav.*, **41**, 334–342.

Rumelhart, D.E. and McClelland, J.L., Eds. (1985). *Parallel Distributed Processing: Explorations in the Microstructure of Cognition*. Vol. 1: *Foundations*. Cambridge, MA: MIT Press.

Rutter, M. (1991). A fresh look at "maternal deprivation." In *The Development and Integration of Behavior*, ed. P. Bateson. Cambridge, UK: Cambridge University Press, pp. 331–374.

Rutter, M. (2002). Nature, nurture, and development: from evangelism through science toward policy and practice. *Child Develop.*, **73**, 1–21.

Ryan, M.J. (1980). Female mate choice in a Neotropical frog. *Science*, **209**, 523–525.

Ryan, M.J. (1983). Sexual selection and communication in a Neotropical frog, *Physalaemus pustulosus*. *Evolution*, **39**, 261–272.

Ryan, M.J. (1985). *The Túngara Frog, A Study in Sexual Selection and Communication*. Chicago: University of Chicago Press.

Ryan, M.J. (1990). Sensory systems, sexual selection, and sensory exploitation. *Oxford Surv. Evol. Biol.*, **7**, 157–195.

Ryan, M.J. (1996). Phylogenetics and behavior; some cautions and expectations. In *Phylogenetics and Behavior; Some Cautions and Expectations*, ed. E. Martins. Oxford, UK: Oxford University Press. pp. 1–21.

Ryan, M.J. (1998). Receiver biases, sexual selection and the evolution of sex differences. *Science*, **281**, 1999–2003.

Ryan, M.J. (2005). The evolution of behaviour, and integrating it towards a complete and correct understanding of behavioural biology. *Anim Biol.*, **55**, 419–439. (Chapter 7 in this volume.).

Ryan, M.J. and Getz, W. (2000). Signal decoding and receiver evolution: an analysis using an artificial neural network. *Brain, Behav. Evol.*, **56**, 45–62.

Ryan, M.J. and Rand, A.S. (1993). Sexual selection and signal evolution: the ghost of biases past. *Phil. Trans. Roy. Soc. ser. B.*, **340**, 187–195.

Ryan, M.J. and Rand, A.S. (1995). Female responses to ancestral advertisement calls in Tungara frogs. *Science (Washington, DC)*, **269**, 390–392.

Ryan, M.J. and Rand, A.S. (1999). Phylogenetic influences on mating call preferences in female túngara frogs (*Physalaemus pustulosus*). *Anim. Behav.*, **57**, 945–956.

Ryan, M.J. and Rand, A.S. (2003). Mate recognition in túngara frogs: a review of some studies of brain, behavior, and evolution. *Acta Zoologica Sinica.* **49**, 713–726.

Ryan, M.J., Tuttle, M.D., and Rand, A.S. (1982). Sexual advertisement and bat predation in a Neotropical frog. *Am. Nat.*, **119**, 136–139.

Ryan, M.J., Bartholomew, G.A., and Rand, A.S. (1983). Energetics of reproduction in a Neotropical frog, *Physalaemus pustulosus*. *Ecology* **64**, 1456–1462.

Ryan, M.J., Fox, J.H., Wilczynski, W., and Rand, A.S. (1990). Sexual selection for sensory exploitation in the frog *Physalaemus pustulosus*. *Nature*, **343**, 66–67.

Ryan, M.J., Rand, W., Hurd, P.L., Phelps, S.M., and Rand, A.S. (2003). Generalization in response to mate recognition signals. *Am. Nat.*, **161**, 380–394.

Sakata, J.T. and Crews, D. (2004). Developmental sculpting of social phenotype and plasticity. *Neurosci. Biobehav. Rev.*, **28**, 95–112.

Sakata, J.T., Coomber, P., Gonzalez-Lima, F., and Crews, D. (2000). Functional connectivity among limbic brain areas: differential effects of incubation temperature and gonadal sex in the leopard gecko, *Eublepharis macularius*. *Brain, Behav. Evol.*, **55**, 139–151.

Sakata, J.T., Gupta, A., and Crews, D. (2001). Animal models of experiential effects on neural metabolism: plasticity in limbic circuits. In *Neuroplasticity, Development and Steroid Hormone Action*, ed. R. Handa, S. Hayashi, E. Terasawa, and M. Kawata. Boca Raton: CRC Press, pp. 257–272.

Sakata, J.T., Gupta, A., Chuang, C.-P., and Crews, D. (2002). Social experience affects territorial and reproductive behaviours in male leopard geckos, *Eublepharis macularius*. *Anim. Behav.*, **63**, 487–493.

Schlaepfer, M.A., Runge, M.C., and Sherman, P.W. (2002). Ecological and evolutionary traps. *Trends Ecol. Evol.*, **17**, 474–480.

Schoener, T.W. (1971). Theory of feeding strategies. *Ann. Rev. Ecol. Syst.*, **2**, 369–404.

Schutz, F. (1965). Sexuelle Prägung bei Anatiden. *Z. Tierpsychol.*, **22**, 50–103.

Schwabl, H. (1993). Yolk is a source of maternal testosterone for developing birds. *Proc. Nat. Acad. Sci. USA*, **90**, 11446–11450.

Schwabl, H. (1996a). Maternal testosterone in the avian egg enhances postnatal growth. *Comp.Biochem. Phys.* **114**, 271–276.

Schwabl, H., Mock, D.W., and Gieg, J.A. (1997). A hormonal mechanism for parental favouritism. *Nature*, **386**, 231–231.

Searcy, W.A. and Anderson, M. (1986). Sexual selection and the evolution of song. *Ann. Rev. Ecol. Syst.*, **17**, 507–533.

Shatz, C.J. (1992). The developing brain. *Scientific Amer.*, **267**, 60–67.

Shelton, J.R. and Caramazza, A. (1999). Deficits in lexical and semantic processing. Implications for models of normal language. *Psychon. Bull. Rev.*, **6**, 5–27.

Sherman, P.W. (1988). The levels of analysis. *Anim. Behav.*, **36**, 616–619.

Sherman, P.W., Reeve, H.K., and Pfennig, D.W. (1997). Recognition systems. In *Behavioural Ecology: An Evolutionary Approach*, ed. J.R. Krebs and N.B. Davies. 4th edn. Oxford, UK: Blackwell Science, pp. 69–96.

Sherry, D.F. (1985). Food storage by birds and mammals. *Adv. Study Behav.*, **15**, 153–188.

Sherry, D.F. (2005). Do ideas about function help in the study of causation? *Anim. Biol.*, **55**, 441–456. (Chapter 8 in this volume.).

Sherry, D.F. and Schacter, D.L. (1987). The evolution of multiple memory systems. *Psychol. Rev.*, **94**, 439–454.

Sherry, D.F. and Vaccarino, A.L. (1989). Hippocampus and memory for food caches in Black-Capped Chickadees. *Behav. Neurosci.*, **103**, 308–318.

Sherry, D.F., Krebs, J.R., and Cowie, R.J. (1981). Memory for the location of stored food in marsh tits. *Anim. Behav.*, **29**, 1260–1266.

Sherry, D.F., Vaccarino, A.L., Buckenham, K., and Herz, R.S. (1989). The hippocampal complex of food-storing birds. *Brain Behav. Evol.*, **34**, 308–317.

Shettleworth, S.J. (1995). Comparative studies of memory in food storing birds: from the field to the Skinner box. In *Behavioral Brain Research in Naturalistic and Semi-Naturalistic Settings*, ed. E., Alleva, A., Fasolo, H.P., Lipp, L., Nadel, and Ricceri, L. Kluwer Academic, pp. 159–194.

Shettleworth, S.J. (1998). *Cognition, Evolution, and Behavior*. Oxford, UK: Oxford University Press.

Shettleworth, S.J. (2000). Modularity and the evolution of cognition. In *The Evolution of Cognition*, ed. C. Heyes, and L. Huber. Cambridge, MA: MIT Press, pp. 43–60.

Shettleworth, S.J. (2003). Memory and hippocampal specialization in food-storing birds: challenges for research on comparative cognition. *Brain Behav. Evol.*, **62**, 108–116.

Shettleworth, S.J. (2007). Planning for breakfast. *Nature*, **445**, 825–826.

Shettleworth, S.J. and Krebs, J.R. (1982). How marsh tits find their hoards: the roles of site preference and spatial memory. *J. Exp. Psychol.: Anim. Behav. Proc.*, **8**, 354–375.

Shettleworth, S.J. and Westwood, R.P. (2002). Divided attention, memory, and spatial discrimination in food-storing and nonstoring birds, black-capped chickadees (*Poecile atricapilla*). and dark-eyed juncos (*Junco hyemalis*). *J. Exp. Psychol.: Anim. Behav. Proc.*, **28**, 227–241.

Shettleworth, S.J., Hampton, R.R., and Westwood, R.P. (1995). Effects of season and photoperiod on food storing by black-capped chickadees, *Parus atricapillus*. *Anim. Behav.*, **49**, 989–998.

Shine, R. (1979). Sexual selection and sexual dimorphism in the amphibia. *Copeia*, **1979**, 297–306.

Short, T.L. (2002). Darwin's concept of final cause: neither new nor trivial. *Biol. Philos.*, **17**, 323–340.

Silver, R. (1990). Biological timing mechanisms with special emphasis on the parental behavior of doves. In *Contemporary Issues in Comparative Psychology*, ed. D.A. Dewsbury. Sunderland, MA: Sinauer, pp. 252–277.

Singer, A.G., Beauchamp, G.K., and Yamazaki, K. (1997). Volatile signals of the major histocompatibility complex in male mouse urine. *Proc. Natl Acad. Sci. USA*, **94**, 2210–2214.

Skinner, B.F. (1953). *Science and Human Behavior*. New York: Macmillan.

Slater, P.J.B. (1999). *Essentials of Animal Behaviour.* Cambridge, UK: Cambridge University Press.

Slater, P.J.B., Eales, L.A., and Clayton, N.S. (1988). Song learning in zebra finches: progress and prospects. *Adv. Study Behav.*, **18**, 1–34.

Sluckin, W. and Salzen, E.A. (1961). Imprinting and perceptual learning. *Q. J. Exp. Psychol.*, **8**, 65–77.

Smulders, T.V. and DeVoogd, T.J. (2000). The avian hippocampal formation and memory for hoarded food: spatial learning out in the real world. In *Brain, Perception, Memory*, ed. J.J. Bolhuis. Oxford, UK: Oxford University Press, pp. 127–148.

Smulders, T.V., Sasson, A.D., and DeVoogd, T.J. (1993). Seasonal changes in brain size in a food-storing bird, the black-capped chickadee (*Parus atricapillus*). *Soc. Neurosci. Abstr.*, **19**, 1448.

Smulders, T.V., Sasson, A.D., and DeVoogd, T.J. (1995). Seasonal variation in hippocampal volume in a food-storing bird, the black-capped chickadee. *J. Neurobiol.*, **27**, 15–25.

Smulders, T.V., Shiflett, M.W., Sperling, A.J., and DeVoogd, T.J. (2000). Seasonal change in neuron number in the hippocampal formation of a food-hoarding bird: the black-capped chickadee. *J. Neurobiol.*, **44**, 414–422.

Snowdon, C.T. and Hausberger, M. (eds.) (1997). *Social Influences on Vocal Development.* Cambridge, UK: Cambridge Univ. Press.

Sober, E. (1984). *The Nature of Selection.* Cambridge, MA: MIT Press.

Sockman, K.W., Gentner, T.Q., and Ball, G.F. (2002). Recent experience modulates forebrain gene-expression in response to mate-choice cues in European starlings. *Proc. Roy. Soc. Lond., B.*, **269**, 2479–2485.

Sokolowski, M.B. (2001). Drosophila: genetics meets behaviour. *Nature Rev. Genet.*, **2**, 879–890.

Solis, M.M., Brainard, M.S., Hessler, N.A., and Doupe, A.J. (2000). Song selectivity and sensorimotor signals in vocal learning and production. *Proc. Natl. Acad. Sci. USA*, **97**, 11836–11842.

Steer, M. and Cuthill, I.C. (2003). Irrationality, sub-optimality and the evolutionary context. *Behav. Brain Sci.*, **26**, 176–177.

Stephens, D.W. and Krebs, J.R. (1986). *Foraging Theory.* Princeton NJ: Princeton University Press.

Stevenson, J.G. (1967). Reinforcing effects of chaffinch song. *Anim. Behav.*, **15**, 427–432.

Suthers, R.A. (2001). Peripheral vocal mechanisms in birds: are songbirds special? *Neth. J. Zool*, **51**, 217–242.

ten Cate, C. (1984). The influence of social relations on the development of species recognition in zebra finch males. *Behaviour*, **91**, 263–285.

ten Cate, C. (1987). Sexual preferences in zebra finch males raised by two species: II. The internal representation resulting from double imprinting. *Anim. Behav.*, **35**, 321–330.

ten Cate, C. (1989). Behavioural development: toward understanding processes. In *Perspectives in Ethology, Vol 8.* ed. P.P.G. Bateson and P.H. Klopfer. New York: Plenum, pp 243–269.

ten Cate, C. (1994). Perceptual mechanisms in imprinting and song learning. In *Causal Mechanisms of Behavioural Development*, ed. J.A. Hogan and J.J. Bolhuis. Cambridge, UK: Cambridge University Press, pp. 116–146.

ten Cate, C. and Bateson, P. (1988). Sexual selection: the evolution of conspicuous characteristics in birds by means of imprinting. *Evolution*, **42**, 1355–1358.

ten Cate, C. and Vos, D.R. (1999). Sexual imprinting and evolutionary processes in birds: a reassessment. *Adv. Study Behav.*, **28**, 1–31.

Terpstra, N.J., Bolhuis, J.J., and den Boer-Visser, A.M. (2004). An analysis of the neural representation of bird song memory. *J. Neurosci.* **24**, 4971–4977.

Terpstra, N.J., Bolhuis, J.J., Riebel, K., van der Burg, J.M.M., and den Boer-Visser, A.M. (2006). Localised brain activation specific to auditory memory in a female songbird. *J. Comp.Neurol.*, **494**, 784–791.

Thornhill, R. (1990). The study of adaptation. In *Interpretation and Explanation in the Study of Behavior*, ed. M. Bekoff and D. Jamieson. Boulder, CO: Westview Press, pp. 1–31.

Thorpe, W.H. (1961). *Bird Song*. Cambridge, UK: Cambridge University Press.

Timmermans, S., Lefebvre, L., Boire, D., and Basu, P. (2000). Relative size of the hyperstriatum ventrale is the best predictor of feeding innovation rate in birds. *Brain Behav. Evol.*, **56**, 196–203.

Tinbergen, N. (1951). *The Study of Instinct*. Oxford, UK: Oxford University Press.

Tinbergen, N. (1952). Derived activities: their causation, biological significance, origin and emancipation during evolution. *Q. Rev. Biol.*, **27**, 1–32.

Tinbergen, N. (1953). *Social Behaviour in Animals*. London: Methuen.

Tinbergen, N. (1963). On aims and methods of ethology. *Z. Tierpsychol.*, **20**, 410–433. (Reprinted in this volume.)

Tinbergen, N. (1985). Watching and wondering. In *Leaders in the Study of Animal Behavior: Autobiographical Perspectives*, ed. D.A. Dewsbury. Chapter 17. Lewisburg, PA: Bucknell University Press.

Tinbergen, N. and Kruyt, W. (1938). Über die Orientierung des Bienenwolfes (*Philanthus trangulum* Fabr.). III: Die Bevorzugung bestimmter Wegmarken. *Z. Vgl. Physiol.*, **25**, 292–334.

Tinbergen, N., Brockhuysen, G.J., Feekes, F., Houghton, J.C.W., Kruuk, H., and Szulc, E. (1962). Egg shell removal by the black headed gull, *Larus ridibundus*: a behaviour component of camouflage. *Behaviour*, **19**, 74–117.

Toates, F. (1986). *Motivational Systems*. London: Cambridge University Press.

Toates, F. and Jensen, P. (1991). Ethological and psychological models of motivation: towards a synthesis. In *From Animals to Animats*. Cambridge, MA: MIT Press, ed. J.A. Meyer and S. Wilson, pp. 194–205.

Tomasello, M. (1995). Language is not an instinct. *Cogn. Devel.*, **10**, 131–156.

Tomasello M. (2000). Primate cognition: introduction to the issue. *Cogn. Sci.*, **24**, 351–361.

Tramontin, A.D. and Brenowitz, E.A. (1999). A field study of seasonal neuronal incorporation into the song control system of a songbird that lacks adult song learning. *J. Neurobiol.* **40**, 316–326.

Tramontin, A.D. and Brenowitz, E. (2000). Seasonal plasticity in the adult brain. *Trends Neurosci.*, **23**, 251–258.

Trivers, R.L. (1971). The evolution of reciprocal altruism. *Q. Rev. Biol.*, **46**, 35–57.

Trivers, R.L. (1972). Parental investment and sexual selection. In *Sexual Selection and the Descent of Man, 1871–1971*, ed. B.G. Campbell. Chicago: Aldine, pp. 136–179.

Trivers, R.L. (1974). Parent–offspring conflict. *Amer. Zool.*, **14**, 249–264.

Trivers, R.L. and Hare, H. (1976). Haplodiploidy and the evolution of the social insects. *Science*, **191**, 249–263.

Trivers, R.L. and Willard, D.E. (1973). Natural selection of parental ability to vary? the sex ratio of offspring. *Science*, **179**, 90–92.

Tulving, E. (1983). *Elements of Episodic Memory*. New York: Oxford University Press.

Tuttle, M.D. and Ryan, M.J. (1981). Bat predation and the evolution of frog vocalizations in the Neotropics. *Science*, **214**, 677–678.

van der Klaauw, C.J. (1940). Theoretische biologie. *Vakbl. Biol.*, **21**, 75–88.

Vander Wall, S.B. (1982). An experimental analysis of cache recovery in Clark's nutcracker. *Anim. Behav.*, **30**, 84–94.

van Dierendonck, M.C. (2006). The importance of social relationships in horses. Ph.D. thesis Utrecht University, Faculty of Veterinary Medicine, the Netherlands.

van Kampen, H.S. (1996). A framework for the study of filial imprinting and the development of attachment. *Psychon. Bull. Rev.*, **3**, 3–20.

Ventura, D.F., Zana, Y., De Souza, J.M., and Devoe, R.D. (2001). Ultraviolet colour opponency in the turtle retina. *J. Exp. Biol.*, **204**, 2527–2534.

Vestergaard, K.S. (1982). Dust-bathing in the domestic fowl: diurnal rhythm and dust deprivation. *Appl. Anim. Ethol.*, **8**, 487–495.

Vestergaard, K.S. (1994). Dustbathing and its relation to feather pecking in the fowl: motivational and developmental aspects. Ph.D. thesis, Royal Veterinary and Agricultural University, Copenhagen.

Vestergaard, K.S. and Baranyiova, E. (1996). Pecking and scratching in the development of dust perception in young chicks. *Acta Veter. Brno*, **65**, 133–142.

Vestergaard, K.S. and Hogan, J.A. (1992). The development of a behavior system: dustbathing in the Burmese red junglefowl. III. Effects of experience on stimulus preference. *Behaviour*, **121**, 215–230.

Vestergaard, K.S. and Lisborg, L. (1993). A model of feather pecking development which relates to dustbathing in the fowl. *Behaviour*, **126**, 291–308.

Vestergaard, K.S., Hogan, J.A., and Kruijt, J.P. (1990). The development of a behavior system: dustbathing in the Burmese red junglefowl. I, The influence of the rearing environment on the organization of dustbathing. *Behaviour*, **112**, 99–116.

Vestergaard, K.S., Damm, B.I., Abbott, U.K., and Bildsøe, M. (1999). Regulation of dustbathing in feathered and featherless domestic chicks: the Lorenzian model revisited. *Anim. Behav.*, **58**, 1017–1025.

Viitala, J., Korpimäki, E., Palokangas, P., and Koivula, M. (1995). Attraction of kestrels to vole scent marks visible in ultraviolet light. *Nature*, **373**, 425–427.

Viola, A.U., Archer, S.N., James, L.M. *et al.* (2007). *PER3* polymorphism predicts sleep structure and waking performance. *Current Biol.*, **17**, 613–618.

vom Saal, F., Clark, M., Galef, B., Drickamer, L.C., and Vandenbergh, J.G. (1999). Intrauterine position phenomenon. In *Encyclopedia of Reproduction. Vol. 2*, C. Knobil and J.N. Neill. New York: Academic Press. pp. 893–900.

von Engelhardt, N. (2004). Proximate control of avian sex allocation, a study on zebra finches. Ph.D. thesis. Groningen, the Netherlands.

von Holst, E. and von St. Paul, U. (1960). Vom Wirkungsgefüge der Triebe. *Naturwissenschaft*, **47**, 409–422. (Trans.: On the functional organization of drives *Anim. Behav.*, 1963, **11**, 1–20.).

Walker, M.M., Dennis, T.E., and Krischvink, J.L. (2002). The magnetic sense and its use in long-distance navigation by animals. *Curr. Opin. Neurobiol.*, **12**, 735–744.

Ward, B.C., Nordeen, E.J., and Nordeen, K.W. (1998). Individual variation in neuron number predicts differences in the propensity for avian vocal imitation. *Proc. Natl Acad. Sci. USA*, **95**, 1277–1282.

Weaver, A. and de Waal, F.B.M. (2002). An index of relationship quality based on attachment theory. *J. Comp. Psychol.*, **116**, 93–106.

Weinstock, M. (1997). Does prenatal stress impair coping and regulation of hypothalamic-pituitary-adrenal axis? *Neurosci. Biobehav. Rev.*, **21**, 1–10.

Werker, J.F. and Tees, R.C. (1992). The organization and reorganization of human speech perception. *Ann. Rev. Neurosci.*, **15**, 377–402.

West, M.J. and King, A.P. (1988). Female visual display affects the development of males' song in the cowbird. *Nature*, **334**, 244–246.

West, M.J. and King, A.P. (2001). Science lies its way to the truth … really. In *Handbook of Behavioral Neurobiology*, Vol. 13: *Developmental Psychobiology*, ed. E.M. Blass. New York: Kluwer Academic/Plenum, pp. 587–614.

Westneat, D.F. and Birkhead, T.R. (1998). Alternative hypotheses linking the immune system and mate choice for good genes. *Proc. Roy. Soc. Lond. B*, **265**, 1065–1073.

Whitman, C.O. (1898). *Animal Behavior*. Woods Hole.

Widowski, T.M. and Duncan, I.J.H. (2000). Working for a dustbath: are hens increasing pleasure rather than reducing suffering? *Appl. Anim. Behav. Sci.*, **68**, 39–53.

Wilczynski, W. and Capranica, R.R. (1984). The auditory system of anuran amphibians. *Progr. Nerwobiol.*, **22**, 1–38.

Wilczynski, W., Rand, A.S., and Ryan, M.J. (1995). The processing of spectral ones by the call analysis system of the túngara frog, *Physalaemus pustulosus, Animal Behavior* **49**.

Wilczynski, W., Rand, A.S., and Ryan, M.J. (2001). Evolution of calls and auditory tuning in the *Physalaemus pustulosus* species group. *Brain, Behavior, and Evolution*, **58**, 137–151.

Wilkinson, G.S. and Reillo, P.R. (1994). Female choice response to artificial selection on an exaggerated male trait in a stalk-eyed fly. *Proc. Roy. Soc. Lond. B*, **255**, 1–6.

Williams, G.C. (1966). *Adaptation and Natural Selection*. Princeton, New Jersey: Princeton University Press.

Wilson, E.O. (1975). *Sociobiology. The New Synthesis*. Cambridge, MA: Belknap Press.

Wiltschko, W. and Wiltschko, R. (2002). Magnetic compass orientation in birds and its physiological basis. *Naturwissenschaften*, **89**, 445–452.

Wimberger, P.H. and de Queiroz, A. (1996). Comparing behavioral and morphological characters as indicates of phylogeny. In *Philogenies and the Comparative Method in Animal Behavior*, ed. E.M. Martins. Oxford: Oxford University Press.

Wingfield, J.C., Hegner, R.E., Dufty Jr. A.M., and Ball, G.F. (1990). The Challenge hypothesis: theoretical implications for patterns of testosterone secretion, mating systems, and breeding strategies. *Amer. Nat.*, **136**, 829–846.

Witkin, J.W. (1992). Reproductive history affects the synaptology of the aging gonadotropin-releasing hormone system in the male rat, *J. Neuroendocr.*, **4**, 427–432.

Wolfer, D.P., Litvin, O., Morf, S., Nitsch, R.M., Lipp, H-P., and Würbel, H. (2004). Cage enrichment and mouse behaviour. *Nature*, **432**, 821–822.

Wouters, A.G. (2003). Four notions of biological function. *Studies History Phil. Biol. Biomed. Sci.*, **34**, 633–668.

Wouters, A.G. (2005). The functional perspective of organismal biology. In *Current Themes in Theoretical Biology*, ed. T.A.C. Reydon and L. Hemerik. The Netherlands: Springer, Dordrecht, pp. 33–69.

Würbel, H. (2001). Ideal homes? Housing effects on rodent brain and behaviour. *Trends Neurosci.*, **24**, 207–211.

Yamazaki, K., Boyse, E.A., Miké, V. et al. (1976). Control of mating preferences in mice by genes in the major histocompatibility complex. *J. Exp.Med.*, **144**, 1324–1335.

Yamazaki, K., Beauchamp, G.K., Curran, M., Bard, J., and Boyse, E.A. (2000). Parent–progeny recognition as a function of MHC odortype identity. *Proc. Natl Acad. Sci. USA*, **97**, 10500–10502.

Yokoyama, S. (1999). Molecular bases of color vision in vertebrates. *Genes Genet. Syst.*, **74**, 189–199.

Yokoyama, S. and Shi, Y.S. (2000). Genetics and evolution of ultraviolet vision in vertebrates. *FEBS Lett.*, **486**, 167–172.

Zana, Y., Ventura, D.F., de Souza, J.M., and Devoe, R.D. (2001). Tetrachromatic input to turtle horizontal cells. *Vis. Neurosci.*, **18**, 759–765.

Zielinski, W.J., vom Saal, F.S., and Vandenbergh, J.G. (1992). The effect of intrauterine position on the survival, reproduction and home range size of female house mice. *Behav. Ecol. Sociobiol.*, **30**, 185–191.

# Index

Page entries for headings with subheadings refer to general aspects of that topic.
Page entries in **bold** refer to figures and diagrams.

acquired behavior patterns xi
action specific energy xviii
activation energy 52–53
*Adaptation and Natural Selection*
  (Hamilton & Williams) 129
adaptationist stance 108, 112, 166,
  167; *see also* function
adaptations
  and aptations 110–111
  ontogenic 56, 103
adaptive specializations approach,
  critique of 167, 170
adult development 18; *see also*
  postnatal environment
age effects, leopard gecko 76
aggressive behavior
  development of individual
    differences 60
  mice 77, **78**
  sticklebacks 15, xv–xvi
allogrooming in horses 46
altruism, reciprocal 129
American experimental psychology ix
amphibians, mating calls 20,
  149; *see also* túngara frog
analytical phase of ethology 3
androgens, in birds 68–69, 79–80
  between clutch variation 73–74
  long-lasting effects 74–75
  within clutch variation 69–73, **69**, **71**
androgen to estrogen ratios,
  geckos 60
*Animal Minds* (Griffin) 40
animals; *see also* birds
  cognition *see* animal cognition

integrative approach to animal
  behavior 127–128, 131
social relationships 46
welfare 44–46, xix–xx
anthropocentric program 164
anthropomorphism 36, 40, 43
applied ethology xix, 43–46; *see also*
  behavioral biology; ethology
  animal welfare 44–46
  dustbathing in poultry 45–46
  expression of normal behavior 44–46
  feather pecking in poultry 43–44
  pleasure, animal 44–46
  social relationships in horses/
    allogrooming 46
aptations/adaptations 110–111
architectural metaphor xiv
Aristotle 28
*Art of the Soluble* (Medawar) xii
attachment behavior 101–102
auditory
  localization, barn owls 157–158
  templates, bird song 91–92
avian species *see* birds

babbling, infant 93, 98
barn owls, auditory localization
  157–158
bees 9, 14, 119–120
behavior genetics xv–xviii
behavior mechanisms 36–37,
  193–195
  developmental changes 55,
    82; *see also* development;
    ontogeny

hierarchy of 6
integration with function
    166–167; *see also* function;
    survival value
modularity 105–106
perceptual/central/motor 36–38
releasing mechanisms (RM) x, 5,
    10–11, 31–32, 36, 47–48
behavior systems **37**, 94–97
contentment mechanism 86, 87
dustbathing, fowl 95–97, **96**
filial imprinting 85–86, **86**
human language as 101
hunger system, junglefowl 94–95,
    **95**
levels of analysis 37–38, 39
behavior theory critique, Lehrman
    x–xi
behavioral
    biology xxi, 1, 2, 25; *see also*
        behavioral ecology; ethology;
        four questions
    ecology *see below*
    endocrinology 39
    genetics 39
    syndromes (personality) 30
behavioral ecology xiii, 113–115,
    122–125
comparative method 120–122
critiques 124–125
design approach 123–124
developmental/genetic
    constraints 117
experimental manipulations
    117–118
field experiments 118
phenotype manipulation
    experiments 117, 124
mathematical modeling 114, **115**,
    118–120, 125
mechanisms 123–124
observational studies
    115–117, 125
optimality 114, 124
optimization model 117
phylogenetically independent
    contrasts method 121–122
selection experiment 117
and sociobiology 122–123
Tinbergen's methodology
    114–115, **115**
biological method 2, 25
biological study of behavior,
    ethology as xxi, 2, 17,
    20, 25
biology, behavioral xxi, 1, 2, 25; *see also*
    behavioral ecology; ethology;
    four questions

birds; *see also* fowl
    auditory localization, owls 157–158
    beaks xiv, 10
    clutch variation *see* androgens
    domestic *see* fowl
    dunnocks (*Prunella modularis*)
        xvii–xviii
    eggshell removal behavior, gulls
        xiv, 13–14
    feeding behavior xiii, 10
    food-caching 41–42, 153–154,
        187–188; *see also* hippocampal
        hypothesis; neurobiology
    kittiwake behavior 2, 12–13, 20
    postnatal influences on
        development 77–80
    pre-hatching experiences, maternal
        calls 87–88
    sexual imprinting 89–91
    social interactions 79
    song *see below*
    ultraviolet vision 110–111
bird song xiv, xvi, 128
    auditory templates 91–92
    and development 15, 91–94, 103
    conspecific song 92
    and four questions 26
    hippocampal neurogenesis 155
    memorization 92, 179–181; *see also*
        song systems
    neurobiological studies 103, 128
    seasonal brain variations 181–183
    song systems 168–170, **169**,
        172, 176
    stages in song production 93
    synapse density 180
    black-headed gulls, eggshell
        removal behavior xiv, 13–14
blackbird's bill 10
brain modularity 175–177; *see also*
    hippocampal hypothesis;
    neurobiology/neuroecology
brain nuclei
    metabolic activity 66
    reorganization of functional
        associations 76
brain regions, localization 168–170,
    **169**; *see also* hippocampal
    hypothesis; neurobiology/
    neurecology; song systems

caching of food 41–42, 153–154,
    187–188; *see also* hippocampal
    hypothesis; neurobiology
calls, mating 20, 149; *see also* bird song;
    song; túngara frog
camouflage 11–12
caterpillars, crowding 9

cats, sexual experience 75
caudomedial mesopallium (CMM)
    180–181
caudomedial nidopallium (NCM)
    180–181
causal criterion 83
causal models 46
    dustbathing in fowl 48–49
    neural network models 46–48
    psycho-hydraulic model of
        motivation 50–52, **51**
    two process model of human sleep
        49–50
causal questions 163, 165; *see also*
    causation; development
causation xii, 35, 52–53,
        xv–xviii; *see also* applied ethology;
        causal models; cognition (animal)
    activation (energy) 52–53
    complexity 4, 5–6, 38
    conception of 35–38
    definitions 27–28
    and function *see below*
    functional units of behavior 36, 38
    Huxley on xxii, 2, 54, 108, 147
    internal causes of behavior 4, 6, 38
    levels of analysis 36–38, 52–53
    organs of behavior 4–5, 8, 19, 35
    proximate 29–30, 35, 173
    structure (threshold) 52–53
    studies of 38–40, **40**
    and survival value 8
    Tinbergen's conceptualisation 4–6
    ultimate 29–30, 173; *see also*
        function; survival value
causation/function, distinctions/
        synthesis 30, 147–150, 161; *see also*
        auditory localization; magnetic
        field orientation; major
        histocompatibility complex;
        memory in food-storing birds
    behavior patterns as organs 149
    causal/functional distinctions
        147–148, 165
    causal/functional synthesis 150,
        160–161
    criteria for causal explanations 157
    discrete functional/causal units,
        non-equivalence 150
    function as clue to causation 150
    hammer example 147
    human foresight 148
    interaction of four questions 147
    limitations/hazards of functional
        approach 149–150
    natural selection 148
    same functions having different
        causes 149

cause 27–28; *see also* causation
chaffinches, song development 16
challenge hypothesis 79–80
chickens *see* fowl
circadian rhythms
    dustbathing in fowl 49, 51
    uncoupling 49
cliff-nesting, kittiwake behavior 2,
    13, 20
CMM (caudomedial mesopallium)
    180–181
cognition, animal 40–43
    definition of cognition 41
    episodic memory 41–42
    jumping spiders (*Portia fimbriata*)
        42–43
    primate studies 41
    scrub jays, food-caching behavior
        41–42
cognitive ecology 163, 166, 167
*Cognitive Ethology; the minds of other
    animals* (Ristau) xx
color mutations 14
communal nesting/nursing 152
communication, sexual *see* bird
    song; mating calls; song;
    túngara frog
comparative
    method 120–122
    psychologists 27
"compass," animal 159
competitive exclusion model 84, 85
complexity, behavioral 4, 5–6, 16, 38
computational question 112
conflict hypothesis xviii
consciousness, animal xix–xx; *see also*
    cognition (animal)
contentment mechanism 86, 87
controls, experimental 10
corticosterone levels, food-hoarding
    species 186–187
COX (cytochrome oxidase) activity
    62–64, **65–66**
crickets, song xvi
critical (sensitive) periods 84–85, 106
"Critique of Konrad Lorenz's
    Theory of Instinctive Behavior"
    54, x–xi
crossbill (*Loxia curvirostra*), beaks xiv
crowding in *Vanessa* caterpillars 9
current utility 109–111, 112–113, 116,
    117; *see also* survival value
cytochrome oxidase (COX) activity
    62–64, **65–66**

dance, bee 9, 14
Darwinism 108
deaf children/signing 101

deep synthesis 164; *see also* integration
definitions/terminology 6, 21
    behavioral biology (ethology) xxi, 1,
        2, 25
    biological study of behavior,
        ethology as xxi, 2, 18, 21, 25
    cause/causation 27–28, 35–38
    causal/functional distinctions
        147–148
    cognition 41
    current utility 109–111, 112–113
    development 55, 82
    dictionary vs. scientific 29–30, 31
    evolution 32
    four questions, integration/
        distinctions 163–164, 165–166
    function 28, 109–111; and *see below*
    historical definitions of function
        110–111, 112–113, 114
    innate behavior 16–18, 104–105
    levels 31
    and nature–nurture debate
        104–105
    neuroecology/neuroethology,
        distinctions 172–173
    neurophysiology/ethology
        boundaries 6–7
    ontogeny 14
    physiology of behavior 6, 31, 38
    survival value 28, 109, 113
    ultimate 31
descriptive/observational phase of
    ethology 2–4
development, post-Tinbergen
    conceptualizations xii, 82–83,
        102–106, xvii–xix; *see also*
        attachment behavior; behavior
        systems; bird song learning; filial
        imprinting; four questions;
        human development; maternal
        environment; ontogeny;
        postnatal environment; sexual
        imprinting
    causal criterion 83
    as change in behavior machinery
        55, 82
    developmental perspective xi,
        xii, xiv
    functional criterion 83
    interactionist interpretation
        83–84, 106
    irreversibility 84, 85, 106
    modularity 105–106
    nature–nurture debate 83–84,
        104–105
    ontogenic adaptations 103
    oversimplifications 84
    predispositions 104–106

psychological insights/cross-
    fertilization 104
    questions 163, 165
developmental neurobiology xvi–xvii
    bird song learning 103
    filial imprinting **65**, 103
displacement activities 6
domain-specific modules 174–175
drives xviii–xix
duck study, sexual imprinting 89
dunnocks (*Prunella modularis*) xvii–xviii
dustbathing in fowl 45–46, 48–49
    as behavior system 95–97, **96**
    circadian rhythms 49, 51

ecological
    approach 166
    program 164
ecology
    behavioral *see* behavioral ecology
    cognitive 163, 166, 167
    functional 163
eggs
    clutch variation *see* androgens
    incubation temperature
        *see* temperature-dependent
        sexual determination
eggshell removal behavior, black-
    headed gulls xiv, 13–14
ejaculation in rats 17
embryonic environment 57; *see also*
    maternal environment;
    temperature-dependent sexual
    determination (TSD)
endocrinology, behavioral 39
engineering, and behavioral
    ecology 124
"English-speaking ethologists" x
episodic memory 41–42
estrogen
    to androgen ratios 60
    to testosterone ratios 66
ethics of research 102
ethologists 27
    "English-speaking" x
    "German group" x
ethology, classical ix; *see also* applied
    ethology; behavioral ecology;
    four questions; *On aims and
    methods of ethology*
    analytical phase 3
    as biological study of behavior xxi,
        2, 18, 21, 25
    descriptive/observational phase 2–4
    father of (Lorenz) 1, 2, 21
    modern perspectives xii
    and neurophysiology 6–7
    rise of 26–27

European school (Lorenz) 54
evolution 17-19, 32, 127, 146; *see also*
    ontogeny; phylogeny; túngara
    frog
  aims of evolutionary study 19
  behavioral xxii, 2, 54, 108, 147
  comparison of closely related
    species 19
  definition 32
  difficulties of studying 17
  dynamics 19
  and four questions xii, xiii,
    xiv-xvii
  Huxley on xxii, 2, 54, 108, 147
  influence of selection on behavior
    evolution 19-20
  integrative approach 127-128, 131
  role 173
  specific examples 128-131
  trait/preference evolution
    138-139, 146
*Evolution, function, development, causation*
    symposium xxi
evolutionary
  biologists 27
  psychology xiii, 163, 166, 167
evolutionarily stable strategy
    concept xiii
experimental psychology,
    American ix
experimental work
  controls 10
  ethics 102
  methodology 19, 117-118
  observational studies 2-4,
    115-117, 125
  practical difficulties 10
  survival value 8-9
eye spots in moths 9

face/facial
  expressions, humans/primates 129
  recognition, humans 87
fanning, male sticklebacks 9
farming 44; *see also* applied ethology
feather pecking, poultry 43-44; *see also*
    applied ethology; fowl
feeding behavior, birds xiii
field experiments 118
fighting behavior in sticklebacks 15,
    xv-xvi
filial imprinting 84-88, **86**, 103
  behavior system concept 85-86, **86**
  competitive exclusion model
    84, 85
  contentment mechanism 86, 87
  and face recognition in humans 87
  neural network model 85

neurobiological studies 103
  and operant conditioning/
    reinforcement 88
  pre-hatching experiences, maternal
    calls, 87-88
  predispositions 86-87, 106
  sensitive/critical periods 84-85, 106
finch studies, sexual imprinting
    89-91
fish 9, 15, 152, xv-xvi
fitness effects 112-113
flower color xiv
food-caching 41-42, 153-154,
    187-188; *see also* hippocampal
    hypothesis; neurobiology
forced desynchrony protocol 49
foresight, human 148
four questions on behavior ix, 25-26,
    33-34, xi-xii, xxi-xxii; *see also*
    causation; development;
    evolution; function; ontogeny;
    survival value
  cause 27-28
  definition problems 27-28, 29-30,
    31, 32
  distinction of questions 165
  formulation of questions 26-27
  interaction of questions 147
  integration of questions 163-164,
    165-166
  levels of analysis issue 31-32
  proximate causation 29-30,
    35; *see also* causation
  ultimate causation 29-30; *see also*
    phylogeny; survival value
fowl; *see also* dustbathing
  feather pecking 43-44; *see also*
    applied ethology
  human face preference in
    chickens 47
  hunger system, junglefowl 94-95,
    **95**-95
  pecking movements, newly
    hatched chicks x
frogs, mating calls 20, 149; *see also*
    túngara frog
fruit fly (*Drosophila melanogaster*) 20, 39
function xii, 107, xii-xiv,
    xiv-xvii; *see also* behavioral
    ecology; causation/function
    (distinctions/synthesis); survival
    value
  aptations/adaptations 110-111
  adaptationist stance 108, 112
  current utility definition 109-111,
    112-113
  definitions, multiple 28, 109-111
  fitness effects 112-113

historical 110-111, 112-113,
    114, 118
Huxley's three divisions 108
ideal free theory 114
integration of function and
    mechanism 166-167
marginal value theorem 114
and neuroecology 163-164,
    195-196; see also neuroecology
philosophical conceptions
    111-113
proximate-ultimate dichotomy
    109, 113
as survival value 28, 107, 109, 113
teleology 111, 166
Tinbergen's conception 107-109,
    165-166
ultraviolet vision in birds 110-111
weak/strong functions 109
functional
    approach 166, 167
    criterion of development 83
    ecology 163
    hypothesis xii, xii-xiv
    questions 163, 165
    units of behavior 36, 38

game theory 114
gecko, social experience 75-76; see also
    temperature-dependent sexual
    determination
gender roles of girls 67
generative grammar 100
genetics
    behavioral 39
    molecular xvi
    sexual determination 57-58
    and species-specific behavior 19
"German group" of ethologists x
grammar 100
grooming
    allogrooming in horses 46

hammer example of causation and
    function 147
haplodiploidy 129
"hard core" x
Heidegger, Martin 147
helpers at the nest xiv
hierarchy of behavior mechanisms 6
hippocampal hypothesis, food-
    hoarding species 153-154, 170,
    171-172, 175-177; see also
    neurobiology; spatial
    information storage
    corticosterone levels 186-187
    Eurasian/N. American bird
        comparisons 154-155

neurogenesis 155-157
size/volume of hippocampus
    154-155, 183-187, **184**, 192-193
seasonal variation 156, 188-192
historical contingency hypothesis
    143-144
homeostatic processes, human sleep 49
hormones 6, 39, 66, 67, 79-80; see also
    androgens; maternal
    environment
horses, allogrooming 46
"how" questions 163, 165
human/s
    development 106
    face recognition 87
    facial expressions 129
    foresight 148
    language see language
    neuronal stem cell therapy 157
    research ethics 102
    sleep 39, 49-50
Hume, David 194
hunger system, junglefowl 94-95, **95**
Huxley, Julian xxii, 2, 25, 54, 108,
    147; see also four questions
hyena 67

ideal free theory 114
imprinting see filial imprinting; sexual
    imprinting
inbreeding avoidance 152
incubation temperature, eggs
        see temperature-dependent
        sexual determination
independent contrast method 143
infants
    babbling 93, 98
    developmental influences
        see postnatal environment
information storage, spatial 41-42,
    153-154, 187-188; see also
    hippocampal hypothesis;
    neurobiology
innate
    behavior patterns x, xi,
        15-17; see also internal causes of
        behavior
    definitions 14-16, 104-105
    modules 174
    releasing mechanisms (RM) x, 5,
        31-32, 36, 47-48
insects
    bees 9, 14, 119-120
    caterpillars, crowding 9
    crickets, song xvi
    fruit fly 20, 39
    moths 7, 9
    sociality 129

*Instinkt–Dressur–Verschräkung* 16
instinctive drives xviii–xix
integration
    animal behavior 127–128, 131
    causal/functional synthesis 9,
        160–161
    four questions 163–164, 165–166
    function and mechanism
        166–167
internal causes of behavior 4, 6,
    14–16, 38; *see also innate behavior
    patterns*

jays
    feeding on caterpillars 11
    food-caching behavior
        41–42; *see also* hippocampal
        hypothesis
jumping spiders (*Portia fimbriata*)
    42–43
junglefowl, hunger 94–95, **95**

kaspar-hauser/isolation experiments
    xvi, 79
kin recognition 150–151, 152
kittiwake behavior 2, 12–13, 20

language, human 87, 97–101
    as behavior system 101
    components 100
    deaf children/signing 101
    grammar 100
    infant babbling 93, 98
    morphemes 99
    phonemes 98–99
    semantic systems 99–100
    syntax 100
learning, within-event 88; *see also*
    memory
Lehrman, D. S. 54–55, 82, 83, x–xi
leopard gecko, social experience
    75–76; *see also* temperature-
    dependent sexual
    determination
levels of analysis issue 31–32,
    128; *see also* four questions
    behavior systems 37–38, 39
    and causation 36–38, 52–53
    level-adequate concepts 38
    perceptual/central/motor behavior
        mechanisms 36–38
litter composition in mice 77, **78**
Lorenz, Konrad, ix
    behavior theory critique by
        Lehrman x–xi
    European school 54
    as father of ethology 1, 2, 21
    first meeting with Tinbergen xxi

functional criterion of
    development 83
imprinting, sexual 54
influence on Tinbergen 108
nature–nurture debate 83–84

*maestro* ix, xx, xxii, 165–166; *see also*
    Tinbergen
magnetic field orientation 158–160
magnetic sensory reception 160
major histocompatibility complex
    (MHC) 150–152
mallard duck study, sexual
    imprinting 89
mammals; *see also* humans; rats
    allogrooming in horses 46
    cats, sexual experience 75
    mice 77, 152
    primate studies 41
    sibling effects 64–67
    spotted hyena 67
    squirrels 16
"map and compass," animal 159
marginal value theorem 114
matched-filter hypothesis 134
mate choice
    individual differences 60, **61**
    neurobiology of 133–136, **135**
    recognition 150–151, 152
    stability of preferences 90–91
mating calls 20, 149; *see also* bird song;
    song; túngara frog
maternal
    calls 87–88
    environment *see below*
    stress 67–68
maternal environment, hormone-
    mediated, 64, 80–81
    androgens in birds 68–69
    hormonal condition during
        pregnancy 67–68
mathematical modeling 114, **115**,
    118–120, 125
mating calls *see bird song*; song
mechanisms, behavior *see* behavior
    mechanisms
memory *see also* neurobiology
    bird song 92, 179–181; *see also* bird
        song; song systems
    episodic 41–42
    food-caching 41–42, 153–154,
        187–188; *see also* hippocampal
        hypothesis; neurobiology
methodology, Tinbergen's 114–115,
    **115**
MHC (major histocompatibility
    complex) 150–152
mice 77, 152

modeling, mathematical 114, **115,**
    118–120, 125
modularity 105–106, 173–174
molecular
    biology 6
    genetics xvi
morphemes 99
moths 7, 9
motivation xviii–xix; *see also* psycho-
    hydraulic model
mutations, and differential survival 13

natural selection 148
naturalistic fallacy 194
naturalists, role of 4
nature/nurture debate 55–56, 79, 81,
    83–84, 104–105; *see also* maternal
    environment; postnatal
    environment; temperature-
    dependent sexual determination
    (TSD)
NCM (caudomedial nidopallium)
    180–181
needs, ethological 44–46; *see also*
    welfare
nesting
    communal 152
    kittiwake behavior 2, 12–13, 20
    rats 16
neural networks 46–48
    artificial 144–146
    filial imprinting 85
    releasing mechanism as 47–48
neural phenotypes 61
neurobiology/neurecology 163, 166,
    167, 178; *see also* hippocampal
    hypothesis
    adaptive specializations approach,
        critique of 167, 170
    behavioral/neural correlations,
        causal links 171–172
    bird song learning 103,
        179–181; *see also* song systems
    current debate 192–195
    developmental xvi–xvii
    filial imprinting 103
    functional approach, critique
        167–168
    limitations of 193–195
    localization of brain regions
        168–170, **169**
    of mate preference 133–136, **135**
    misunderstandings/confusions
        172–178
    and neuroethology, distinctions
        172–173
    spatial memory tasks, failure of
        behavioral predictions 170–171

neuroethology 38–39,
    172–173
neurogenesis 155–157
neuronal stem cell therapy 157
neurophysiology/ethology boundaries
    7–8
*Niko's Nature. A life of Tinbergen and His
    Science of Animal Behavior* (Kruuk)
    xi, xxii
*niveau-adäquate terminologie* 6, 38
non-genomic developmental
    influences 56; *see also* maternal
    environment; postnatal
    environment; temperature
    dependent sexual determination
    (TSD)
North American school 54–55
nursing, communal 152
nut cracking in squirrels 16

observational studies 2–4, 115–117,
    125
odor cues in urine 152
offspring/parent conflict 129
*On aims and methods of ethology*
    (Tinbergen) xxi–xxii, 1, 20–22,
    23–24; *see also* causation;
    evolution; ontogeny; survival
    value
    analytical phase of ethology 3
    as biological study of behavior 2,
        18, 21
    causal/functional synthesis 8,
        160–161
    definitions/conceptualizations 1
    descriptive/observational phase of
        ethology 2–4
    Lorenz as father of ethology 1, 2,
        20–21
    terminology 6, 20
ontogeny 14–18, 31, 32, 54–57,
    80–81; *see also* development; four
    questions; nature–nurture debate
    adult development 18
    characterization 15
    conceptualization (Tinbergen) 55, 82
    control of developmental changes
        17–18
    controversy 54–55
    elimination of environmental
        influences 15
    embryonic environment 57; *see also*
        maternal environment;
        temperature dependent sexual
        determination (TSD)
    infant and adult influences
        *see* postnatal environment
    innate behavior 15–17

ontogeny (cont.)
  ontogenic adaptation 56, 103
  proximate/ultimate approaches 80
  sexual/individual differentiation
    54–57
  species-typical behaviors 56
  and survival value 17
operant conditioning/reinforcement
  88
optimal foraging theory xiii, 114, 123
optimality 114, 124
optimization model 117
optomotor response 6
organs of behavior 4–5, 8, 19, 35, 149
oversimplifications 4, 5–6, 55, 84
owls, auditory localization 157–158

parent/offspring conflict 129
parental investment 129
pathogen resistance 152
*Patterns of Behavior: Konrad Lorenz, Niko*
  *Tinbergen and the Founding of*
  *Ethology* (Burkhardt) xi
pecking movements, newly hatched
  chicks x
personality 30
phenotype/s
  manipulation experiments 117, 124
  neural 61
  plasticity 120
phonemes 98–99
*Phylogenetic Systematics* (Hennig) 130
phylogenetic systematics 130; *see also*
  evolution
phylogenetically independent
  contrasts method 121–122
phylogeny 31, 32, 127, 146; *see also*
  *evolution*; túngara frog
physics, and behavioral ecology 124
physiology of behavior 6, 31, 38
placenta 67, 68
plasticity, phenotypic 120
pleasure, animal 44–46
polarity detection 158–160
postnatal environment 75
  avian species 77–80
  sexual experience in rats/cats 75
  social experience in leopard gecko
    75–76
  social experience in mice 77
poultry *see fowl*
predispositions 86–87, 104–106
prefunctionality 104–105
pregnancy, maternal hormonal
  condition 67–68
primate studies 41, 129
problems, solvable and intractable xii,
  xvii, xx

proximate causation 29–30, 31,
  35; *see also* causation
proximate–ultimate dichotomy 80,
  109, 113
psycho-hydraulic model x, 6, 50–52, **51**
psychology
  American experimental ix
  cross-fertilization with biology 104
  evolutionary xiii, 163, 166, 167

questions; *see also* four questions
  causal 163, 165; *see also* causation;
    development
  "how" 163, 165
  "what for" 163, 165

rats
  ejaculation 17
  nest building 16
  sexual behavior 39, 75
reciprocal altruism 129
recognition
  humans 87
  kin 150–151, 152
  mate 150–151, 152
  species 141–142
reinforcement 88
releasing mechanisms (RM) x, 5,
  10–11, 31–32, 36, 47–48
reptiles, social experience
  75–76; *see also* temperature-
  dependent sexual determination
research ethics 102; *see also*
  experimental work
ritualization 129
rocking, of cryptic animals 11–12

scrub jays, food-caching behavior
  41–42; *see also* hippocampal
  hypothesis
selection
  experiments 20, 117
  influence on behavior evolution
    19–20
  sexual 129
semantic systems 99–100
sensitive periods 84–85, 106
sensory exploitation hypothesis
  139–140, **140**, 141
sex ratio evolution 129
sexual
  behavior 39, 75, 77, **78**
  communication *see* bird song;
    mating calls; song; túngara frog
  determination 57–58; *see also*
    temperature dependent sexual
    determination
  differentiation 54–57

dimorphism 62
imprinting xv, 54, 88–91
selection 129
sibling effects in mammals 64–67
signing in deaf children 101
sleep, human 39, 49–50
*Social Behaviour in Animals*
    (Tinbergen) xiii
social behavior xv
    evolution 129
    in horses/allogrooming 46
    insect 129
social experience, influences of
    birds 79
    leopard gecko 75–76
    mice 77, **78**
sociobiology xiii, xiv, 122–123, 129
*Sociobiology* (Wilson) xii, 40
solvable and intractable problems xii,
    xvii, xx
song; *see also* bird song
    crickets xvi
    cross-species attraction to 129
    maternal calls 87–88
song systems 168–170, **169**, 172, 176
Spandrels of San Marco (architectural
    metaphor) xiv
spatial information storage 41–42,
    153–154, 187–188; *see also*
    hippocampal hypothesis;
    neurobiology
species recognition 141–142
species-specificity xv, 5, 19–20
species-typical behaviors 56
spiders (*Portia fimbriata*) 42–43
spotted hyena 67
squirrels 16
stem cell therapy 157
sticklebacks 8, 14, 152, xv–xvi
strange situation test 102
stress, animal 44, 67–68; *see also*
    applied ethology
strong functions 109
*The Study of Instinct* (Tinbergen) ix, xx,
    82, 109–111
subjectivism 4, 36, 40, 43
survival value 7–13, 39–40; *see also*
    function
    blackbird's bill 10
    and causation 9
    current utility 109–111, 112–113,
    116, 117
    differential survival, *Biston betularia*
    14
    experimental work 7–8
    and function 28, 107, 109, 113
    Huxley on xxii, 2, 54, 108, 147
    "insignificant" behavior 12

misfiring behavior 12
    of ontogeny 17
    rocking of cryptic animals 11–12
    similarity/difference comparisons
    12–13
    of species-specific behavior 19–20
    and teleology 8
synapse density 180
syntactic chains
syntax 100
synthesis, deep 164; *see also*
    integration
synthetic approach 166
systematics, phylogenetic 130; *see also*
    evolution

teleology 4–5, 28, 36, 40
    and function 111, 166
    and survival value 8
temperature dependent sexual
    determination, leopard gecko,
    58–59, **59**, 80
    cytochrome oxidase (COX) activity
    62–64, **66**
    development of individual
    differences 59–61
    unitary neuroanatomical
    framework 61–64
templates, behavior 105–106
terminology *see also* definitions
    *Instinkt–Dressur–Verschräkung* 16
    *niveau-adäquate* 6, 38
testosterone 66, 67, 79–80
*The Question of Animal Awareness*
    (Griffin) xix
theory of mind 41
three major problems of biology xxii,
    2, 25, 54, 108, 147; *see also* four
    questions
Tinbergen; *see also* "*On aims and methods
    of ethology*"
    conception of function 107–109,
    165–166
    first meeting with Lorenz xxi
    influence of Lorenz 108
    methodology 114–115, **115**; *see also*
    observational studies
    Tinbergen's approach to function
*Tinbergen's Legacy*
    credits, xxii–xxiii
    structure of text xxii
treefrogs 149
TSD *see* temperature dependent sexual
    determination
túngara frog (*Physalaemus pustulosus*
    spp.), sexual communication 131,
    140–141
    behavior 142–144

túngara frog (cont.)
  calling behavior 131–133
  evolution 136–140, **137**
  historical contingency 143–144
  sensory exploitation hypothesis
    139–140, **140**, 141
  trait/preference evolution
    138–139, 146
  neurobiology of mate preference
    133–136, **135**
  species recognition 141–142
two-process model of sleep 49–50

ultimate causation 29–30, 31,
    173; *see also* evolution; function;
    survival value
ultimate–proximate dichotomy 80,
    109, 113
ultraviolet vision in birds 110–111

unitary neuroanatomical framework
    61–64, **65–66**; *see also*
    neurobiology/neurecology
urine, odor cues 152

vision
  bees 119–120
  birds 110–111

weak/strong functions 109
welfare, animal 44–46, xix–xx; *see also*
    applied ecology
"what for" questions
    163, 165
within-event learning 88

zebra finch studies 89–91
*Zeitschrift* (journal) xi
zoophysiology 166